Rethinking Causality, Complexity and Evidence for the Unique Patient

Rani Lill Anjum • Samantha Copeland
Elena Rocca
Editors

Rethinking Causality, Complexity and Evidence for the Unique Patient

A CauseHealth Resource for Healthcare Professionals and the Clinical Encounter

With contributions from

Brian Broom, Ivor Ralph Edwards,
Karin Mohn Engebretsen, Roger Kerry,
Anna Luise Kirkengen, Tobias Gustum Lindstad,
Matthew Low, Kai Brynjar Hagen, Christine Price

Editors
Rani Lill Anjum
NMBU Centre for Applied
Philosophy of Science
Norwegian University of
Life Sciences, Ås, Norway

Samantha Copeland
Ethics and Philosophy
of Technology Section
Delft University of Technology
Delft, The Netherlands

Elena Rocca
NMBU Centre for Applied
Philosophy of Science
Norwegian University of
Life Sciences, Ås, Norway

ISBN 978-3-030-41238-8   ISBN 978-3-030-41239-5 (eBook)
https://doi.org/10.1007/978-3-030-41239-5

© The Editor(s) (if applicable) and The Author(s) 2020. This book is an open access publication.
Open Access This book is licensed under the terms of the Creative Commons Attribution 4.0 International License (http://creativecommons.org/licenses/by/4.0/), which permits use, sharing, adaptation, distribution and reproduction in any medium or format, as long as you give appropriate credit to the original author(s) and the source, provide a link to the Creative Commons license and indicate if changes were made.
The images or other third party material in this book are included in the book's Creative Commons license, unless indicated otherwise in a credit line to the material. If material is not included in the book's Creative Commons license and your intended use is not permitted by statutory regulation or exceeds the permitted use, you will need to obtain permission directly from the copyright holder.
The use of general descriptive names, registered names, trademarks, service marks, etc. in this publication does not imply, even in the absence of a specific statement, that such names are exempt from the relevant protective laws and regulations and therefore free for general use.
The publisher, the authors, and the editors are safe to assume that the advice and information in this book are believed to be true and accurate at the date of publication. Neither the publisher nor the authors or the editors give a warranty, expressed or implied, with respect to the material contained herein or for any errors or omissions that may have been made. The publisher remains neutral with regard to jurisdictional claims in published maps and institutional affiliations.

This Springer imprint is published by the registered company Springer Nature Switzerland AG.
The registered company address is: Gewerbestrasse 11, 6330 Cham, Switzerland

In memory of Stephen Tyreman

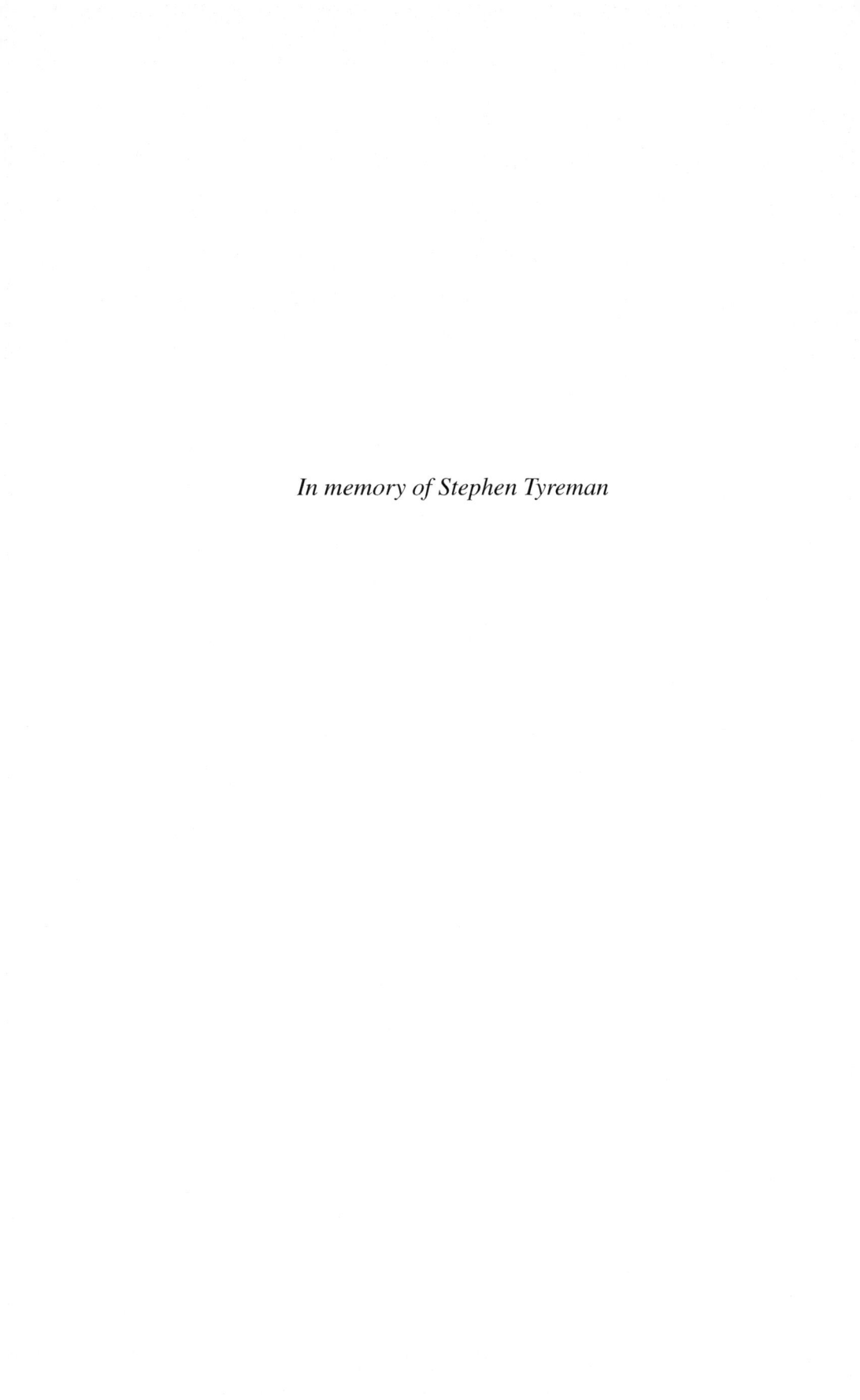

 Norwegian University of Life Sciences
Centre for Applied Philosophy of Science

Preface

The Story of CauseHealth

This book is a result of the interdisciplinary research project CauseHealth, *Causation, Complexity and Evidence in Health Sciences*, at the Norwegian University of Life Sciences (NMBU) from 2015 to 2019. The core team of CauseHealth during those years, located at NMBU, consisted of myself, Stephen Mumford, Samantha Copeland, Elena Rocca and Karin Mohn Engebretsen. Our international network of researchers and practitioners included more than 40 experts from a wide range of specialities: general practice, pharmacology, surgery, oncology, clinical psychology, experimental psychology, psychiatry, physiotherapy, osteopathy, immunology, cardiology, paediatrics, pharmacovigilance, nursing, epidemiology, systems medicine, behavioural medicine, public health, medical sociology, medical ethics, person-centred medicine and practice, evidence-based medicine and practice, medical humanities, medically unexplained symptoms, phenomenology and philosophy of medicine and causation.

The idea for CauseHealth was born back in 2011, after the first meeting of an earlier research project, *Causation in Science*. This project addressed some practical and methodological challenges for establishing causal relationships in science, by tracing these challenges back to how to understand causality, philosophically. Four broad areas were chosen: physics, biology, psychology and social science. Since most of the collaborators were interested in biology, the first project event was on this topic. We held an international conference: Causality and Reductionism in Biology and Beyond. Coincidentally, a few of the talks were on medicine, and in the afterthought of these, the first seed for the CauseHealth project was planted. After all, if it is difficult to understand causality in physics, biology, psychology and social science, then medicine must represent the ultimate challenge. The subject of medicine, the human being, is the unity of them all: the physical, the biological, the mental and the social. Our health affects, and is affected by, all these four dimensions.

There were three themes from the 2011 conference that inspired what later became the CauseHealth project. First, Thor Eirik Eriksen discussed the practical problem of dealing with so-called medically unexplained symptoms. In his presentation 'Waiting for an Explanation', he mentioned how these chronic conditions remain a challenge to the medical profession because of their heterogenic and complex nature (Eriksen et al. 2013). Second, he used the term 'medicalisation', which refers to a reductionist tendency in medicine to define more and more aspects of human life as medical issues. Third, Roger Kerry gave a talk with the title 'Causal Dispositionalism in the Health Sciences' where he argued that randomised controlled trials, the gold standard for establishing causal relationships, have a strong philosophical basis in the neo-Humean difference-making theory of causality. Together, these ideas inspired the three pillars of CauseHealth, which are methodology, ontology and practice:

- The philosophical and conceptual motivation for choice of scientific methodology in medicine (methods)
- The reductionist tendency of treating complex psychosocial phenomena as biomedical ones (ontological model)
- The practical challenge of establishing medical causes in cases of complexity and individual variation (practice)

Initially, the CauseHealth project was directed towards medicine and health sciences in general, without a particular focus on clinical practice. It turned out, however, that the philosophical themes and slogans capturing the essence of the project, such as 'One size does not fit all', 'Statistics don't get me' and 'N=1', struck a chord with clinicians. We therefore wanted to create a resource specifically directed at healthcare practitioners. The plan was to gather the philosophical ideas that are most relevant to the clinical encounter, which can then be used as an intellectual resource for anyone working with individual patients. We have made an effort to present the philosophical material in a way that is accessible for nonphilosophers and to include contributions from eight clinicians and one patient from the CauseHealth network who show a practical way forward that is in line with the philosophical ideas. By including these texts, we hope that the more abstract philosophical ideas will become more concrete and useful for clinical practice.

This book is made possible by the generous funding from the Research Council of Norway and the Uppsala Monitoring Centre, the WHO Collaborating Centre for International Drug Monitoring. As for the contents of the book, this is entirely the result of all those people who have engaged with the CauseHealth project and shared their expertise and ideas with us over these past few years. Without the people who helped me develop and pilot the project from initial idea to proposal – Stephen Mumford, Svein Anders Noer Lie, Thor Eirik Eriksen and Roger Kerry – there wouldn't even have been a CauseHealth project.

CauseHealth has been fuelled by the intellectual power, professional experience and personal engagement of so many. Ever since the beginning of CauseHealth, our collaborators and project network, many of whom have contributed to this book, have been vital sources of insights into the practical concerns and challenges facing clinicians and other healthcare professionals in their daily practice. We have been surprised and overwhelmed by the enthusiasm of practitioners participating in our

CauseHealth events or engaging with us on Twitter. So many of them have been invaluable ambassadors for the project by spreading and translating the philosophical ideas of CauseHealth to colleagues around the globe through talks, blogs, podcasts, articles and social media.

This book is written for and because of them.

Ås, Norway
January 2020

Rani Lill Anjum

Reference

Eriksen TE, Kerry R, Lie SAN et al (2013) At the borders of medical reasoning - aetiological and ontological challenges of medically unexplained symptoms. Philos Ethics Hum Med 8:1747–1753

Contents

Part I Philosophical Framework

1. **Introduction: Why Is Philosophy Relevant for Clinical Practice?** 3
   Rani Lill Anjum, Samantha Copeland, and Elena Rocca

2. **Dispositions and the Unique Patient** 13
   Rani Lill Anjum

3. **Probability for the Clinical Encounter** 37
   Elena Rocca

4. **When a Cause Cannot Be Found** 55
   Rani Lill Anjum and Elena Rocca

5. **Complexity, Reductionism and the Biomedical Model** 75
   Elena Rocca and Rani Lill Anjum

6. **The Guidelines Challenge** 95
   Samantha Copeland

Part II Application to the Clinic

7. **The Complexity of Persistent Pain – A Patient's Perspective** 113
   Christine Price

8. **Above and Beyond Statistical Evidence. Why Stories Matter for Clinical Decisions and Shared Decision Making** 127
   Matthew Low

9. **Causality and Dispositionality in Medical Practice** 137
   Ivor Ralph Edwards

10	**Lessons on Causality from Clinical Encounters with Severely Obese Patients** 149
	Kai Brynjar Hagen
11	**Reflections on the Clinician's Role in the Clinical Encounter** 167
	Karin Mohn Engebretsen
12	**The Relevance of Dispositionalism for Psychotherapy and Psychotherapy Research**................................. 179
	Tobias Gustum Lindstad
13	**Causal Dispositionalism and Evidence Based Healthcare** 201
	Roger Kerry
14	**The Practice of Whole Person-Centred Healthcare**............... 215
	Brian Broom
15	**A Broken Child – A Diseased Woman**......................... 227
	Anna Luise Kirkengen
16	**Conclusion: CauseHealth Recommendations for Making Causal Evidence Clinically Relevant and Informed**..................... 237
	Rani Lill Anjum, Samantha Copeland, and Elena Rocca

Notes on Editors and Contributors

Editors and Contributors to Part I

Rani Lill Anjum is Researcher in Philosophy and Principal Investigator of CauseHealth at the Norwegian University of Life Sciences (NMBU). She leads the NMBU Centre for Applied Philosophy of Science together with Elena Rocca, where she works on the philosophical foundations of science with focus on causality, probability and dispositions. She has written four books with Stephen Mumford: *Getting Causes from Powers*, *Causation: A Very Short Introduction* and *Causation in Science and the Methods of Scientific Discovery*, published with Oxford University Press, and *What Tends to Be: The Philosophy of Dispositional Modality*, with Routledge.

Samantha Copeland is Assistant Professor in the Ethics and Philosophy of Technology section at Delft University of Technology. She was a postdoctoral fellow with the CauseHealth project and is on the editorial board for the annual philosophy thematic of the *Journal of Evaluation in Clinical Practice*. Copeland has published on the ethics of addressing unexpected results in clinical research that involves human subjects, including 'The Case of the Triggered Memory: Serendipitous Discovery and the Ethics of Clinical Research' and 'Unexpected Findings and Promoting Monocausal Claims, a Cautionary Tale'. She is also Coauthor of 'Pharmacovigilance as Scientific Discovery: An Argument for Transdisciplinarity', written with Ralph Edwards and Elena Rocca.

Elena Rocca is an Interdisciplinary Researcher with background in pharmacy, biology and theory of science at the Norwegian University of Life Sciences (NMBU). She is Principal Investigator of the project CauseHealth Risk and Safety and leads the Centre for Applied Philosophy of Science with Rani Anjum. She is on the editorial board of the *International Journal of Risk and Safety in Medicine*. Rocca is interested in the role of philosophical bias in the production and evaluation of evidence. She is Author of 'The Judgements that Evidence-Based Medicine Adopts' and 'Bridging the Boundary Between Scientists and Clinicians' and Coauthor of 'Real or Ideal Risk? Philosophy of Causation Meets Risk Analysis' with Rani Anjum.

Contributors to Part II

Brian Broom was until early 2019 a Clinical Immunologist at Auckland City Hospital and is Adjunct Professor in Psychotherapy at Auckland University of Technology. He is trained in internal medicine and psychotherapy and now works to train clinicians to practice whole-person medicine and healthcare. He has written three books addressing this issue: *Somatic Illness and the Patient's Other Story, Meaning-Full Disease: How Personal Experience and Meanings Cause and Maintain Physical Illness* and *Transforming Clinical Practice Using a MindBody Approach: A Radical Integration*.

Ivor Ralph Edwards is Professor of Medicine and Senior Advisor (Former Director) at the Uppsala Monitoring Centre, the WHO Collaborating Centre for International Drug Monitoring. He has worked in clinical toxicology in the fields of drug abuse, acute and chronic poisoning, toxicity from industrial chemicals as well as adverse drug reactions. He now works on medical and legal aspects of causality evaluation, as well as issues of risk and benefit evaluation and data mining approaches to support signal detection and evaluation. He is Coeditor of *Pharmacovigilance: Critique and Ways Forward* and Author of 'Causality Assessment in Pharmacovigilance: Still a Challenge'.

Karin Mohn Engebretsen is a Gestalt Psychotherapist, with over 15 years of clinical experience from private practice. Her doctoral dissertation is on phenomenology and medically unexplained symptoms, with particular focus on burnout. She has written 'Suffering Without a Medical Diagnosis', 'Naked in the Eyes of the Public' and 'Burned Out or "Just" Depressed?'.

Kai Brynjar Hagen is Senior Consultant in the Regional Centre for Morbid Obesity, North Norway. He is also General Practitioner, District Medical Officer for Communicable Diseases and Advising Senior Consultant in the Norwegian Labour and Welfare Administration (NAV). He is Specialist in Community Medicine and is interested in ecological thinking in medicine, from individual to health policy. He is interested in primary causes of obesity development and factors that contribute to maintain obesity, such as trauma or other stressors.

Roger Kerry is Associate Professor in the Division of Physiotherapy and Rehabilitation Sciences, Faculty of Medicine and Health Sciences, University of Nottingham. He specialises on risks and adverse events of manual therapy, neck pain and headache as well as clinical reasoning. He holds a PhD in Philosophy with the doctoral dissertation *Causation and Evidence-Based Medicine*. He is Author of

'Expanding Our Perspectives on Research in Musculoskeletal Science and Practice' and Coauthor of 'The Form of Causation in Health, Disease and Intervention' and 'Causation and Evidence-Based Practice'.

Anna Luise Kirkengen is Professor of General Practice in the Department of Public Health and Nursing, Norwegian University of Science and Technology (NTNU), and has worked as General Practitioner for 30 years. She specialises on the health impacts of childhood violation and is Author of *The Lived Experience of Violation. How Abused Children Become Unhealthy Adults* and *Inscribed Bodies. Health Impact of Childhood Sexual Abuse*, 'Creating Chronicity' and 'From Wholes to Fragments to Wholes—What Gets Lost in Translation?'.

Tobias Gustum Lindstad has extensive background as a Clinical Psychologist both within secondary public mental healthcare and community-based primary care services, as well as in private practice. His research concerns the relevance of philosophy for psychology, psychotherapy research and mental healthcare. He is the main Editor of the book *Respect for Thought: Jan Smedslund's Legacy for Psychology*, coedited with Erik Stänicke and Jaan Valsiner, under contract with Springer.

Matthew Low is a Consultant Physiotherapist in Musculoskeletal Conditions at the Royal Bournemouth and Christchurch Hospitals NHS Foundation Trust. He is a Visiting Associate at the Orthopaedic Research Institute, Bournemouth University. He also teaches on the topics of clinical reasoning and the management of spinal conditions as well as spinal manipulative physiotherapy. He is Author of 'A Novel Clinical Framework: The Use of Dispositions in Clinical Practice. A Person-Centred Approach' and runs the blog *Perspectives on Physiotherapy*.

Christine Price is affected by neuropathic nerve pain, which she has lived with since an injury in 2008. Recently she started to write, blog and talk about her experiences of living well with pain, directed at both clinicians and patients. She has been invited to sit on research advisory panels and is the first patient representative on the executive board of the UK's Physiotherapy Pain Association. She is Coauthor of 'A Person-Centered and Collaborative Model for Understanding Chronic Pain: Perspectives from a Pain Patient, a Practitioner, and a Philosopher' with Rani Anjum and Matthew Low.

Abbreviations

ADHD	attention deficit hyperactivity disorder
APA	American Psychological Association
AUT	Auckland University of Technology
BMI	body mass index
CBT	cognitive behavioural therapy
CT	computed tomography
DSM	Diagnostic and Statistical Manual of Mental Disorders
EBHC	evidence-based healthcare
EBM	evidence-based medicine
EBP	evidence-based practice
GP	general practice
HP	helicobacter pylori
LBP	lower back pain
MSK	musculoskeletal
MUPS	medically unexplained physical symptoms
MUS	medically unexplained symptoms
OA	osteoarthritis
OCD	obsessive-compulsive disorder
PTSD	posttraumatic stress disorder
RCT	randomised controlled trials
RSSO	Regional Centre for Morbid Obesity
SMT	somatic mutation theory
SSRI	selective serotonin reuptake inhibitors
TOFT	tissue organization field theory
WHO	World Health Organization
WPC	whole person care

List of Figures

Fig. 2.1	The original evidence hierarchy of EBM	16
Fig. 2.2	The vector model of causality	26
Fig. 2.3	Subtractive interference	27
Fig. 2.4	Additive interference	28
Fig. 2.5	Dispositions with different strength of tendency	28
Fig. 2.6	A threshold effect T	29
Fig. 4.1	A vulnerable situation, where R is close to the threshold for illness	60
Fig. 4.2	A robust situation, where R is far from the threshold for illness	61
Fig. 5.1	The hierarchy of science	76
Fig. 5.2	A Venn diagram illustrating a reductionist ontology of wholes and parts	77
Fig. 5.3	A change in methods and practice must start from a change in ontology	83
Fig. 7.1	Dispositions on a good day	117
Fig. 7.2	Dispositions on a difficult day	118
Fig. 7.3	A dynamic threshold for health	120
Fig. 7.4	Threshold on a difficult day	120
Fig. 7.5	Mind map illustrating the complexity of my pain	124
Fig. 7.6	The main contributors and improvers of my pain in January	125
Fig. 7.7	The main contributors and improvers of my pain in August	125
Fig. 12.1	The Medical model	182
Fig. 12.2	The vector model part 1: patient and his/her situational circumstances	195
Fig. 12.3	The vector model part 1: patient, situational circumstances and psychotherapist	196

Fig. 13.1 Randomised controlled trial methodology
The randomisation and allocation processes
create the counterfactual conditions
by which $B(x) - B(y)$ would, under a counterfactual
account of causation, constitute a causal claim.
However, causation is happening in each group,
irrespective of the other group. 205

Part I
Philosophical Framework

Chapter 1
Introduction: Why Is Philosophy Relevant for Clinical Practice?

Rani Lill Anjum, Samantha Copeland, and Elena Rocca

1.1 Why Philosophy?

No practice takes place in a philosophical vacuum and medicine is no exception. Health sciences and healthcare practice are enabled, shaped and restricted by some tacit philosophical assumptions, of which practitioners should be aware. What, for instance, does it mean to say that clinical practice should be based on the best available evidence? What counts as the best evidence? And what is the most relevant evidence for the clinical encounter? Although the scientific evidence is itself largely empirical, many normative aspects of evidence based practice are not, as we will explain. In this sense medicine and health sciences, like all sciences, contain some non-empirical elements. These could be concepts, methods, tools, or what we call 'basic implicit assumptions' or *philosophical bias*. We define philosophical bias as

> Basic Implicit Assumptions in Science about how the world is (ontology), what we can know about it (epistemology), or how science ought to be practiced (norms). (Andersen et al. 2019, p. 1)

They count as biases because they skew the development of hypotheses, the design of experiments, the evaluation of evidence and the interpretation of results in specific directions. How we think the world is (ontology) will be reflected in the way we study it (epistemology) and how we think that science ought to be practiced (norms). In medicine and healthcare, philosophical biases would typically influence

---

R. L. Anjum (✉) · E. Rocca
NMBU Centre for Applied Philosophy of Science, Norwegian University of Life Sciences, Ås, Norway
e-mail: rani.anjum@nmbu.no; elena.rocca@nmbu.no

S. Copeland
Ethics and Philosophy of Technology Section, Delft University of Technology, Delft, The Netherlands
e-mail: s.m.copeland@tudelft.nl

choice of methods (e.g. the evidence hierarchy), norms of practice (e.g. standardised treatment) or scientific framework (e.g. the biomedical model).

In the CauseHealth project, *Causation, Complexity and Evidence in Health Sciences*, we wanted to show how philosophical assumptions motivate particular norms, methods and practices in medicine and healthcare. If we want to see a change in the way medicine and healthcare are practiced, we therefore cannot leave the philosophical foundation on which they are based untouched. Any competing practices will require different methods, norms and philosophical assumptions:

PHILOSOPHICAL ASSUMPTIONS → NORMS → METHODS → PRACTICE

For instance, we might want a healthcare system that acknowledges the patient as an integrative whole, where medical issues must be understood not only on the physiological level, but also within a biographical, social and cultural context. However, if the practice of medicine is premised on Descartes' mind-body divide (what is called *dualism*) then no such integration of the whole person can be achieved. It seems, then, that any genuine and permanent change in practice and methodology will have to be motivated by a change in how we think about the world on the most fundamental level. In the words of osteopath Stephen Tyreman:

> …progress and development is not simply a matter of making new discoveries. It is a complex, multi-faceted process that involves good science, yes, but in the context of prevailing socio-cultural ideas and, most importantly, of an individual's world-view. (Tyreman 2018)

This book offers a guide for rethinking some of these more foundational assumptions, or *world-views*, within medicine and healthcare. Such a foundational rethink seems timely and warranted. Since the introduction of evidence based medicine in the 1990s, there has been an increasing interest in methodological, conceptual and ontological discussions among medical researchers, healthcare professionals and philosophers of medicine. There are emerging movements, such as *Person Centered Medicine and Practice*, the *Campaign for Real Evidence Based Medicine* and the *Critical Physiotherapy Network*, to mention only a few. The historian of science, Thomas Kuhn, saw it as a sign of a paradigm in crisis when its members start participating in philosophical discussions (Kuhn 1962). We should not, however, characterise what we see in medicine and healthcare as a scientific crisis so much as a crisis in the philosophy that grounds it (Anjum 2016; Eriksen et al. 2013), as we now go on to show.

1.2 Philosophical Debates in Medicine and Healthcare

A number of concerns have already been raised in the profession about how medicine is defined and practiced, especially when this affects the clinic. We now present briefly some of the debates that are most relevant for the context of this book: debates about medical models (ontology), scientific methodology (epistemology) and clinical practice.

1.2.1 Debating Models (Ontology)

Beyond the Biomedical Model. The biomedical model of health and illness assumes that all medical conditions should be explained as some physiological abnormality. Conditions lacking such biomedical explanation are then characterised as medically unexplained or psychosomatic (Wade and Halligan 2004). A criticism of this is that health complaints must be seen as more complex, containing biological, social and psychological elements. Even if it were desirable to separate the psychosocial causes of health from the 'medical' ones, it might not even be possible (Arnaudo 2017). The bio-psychosocial model proposed by Engel (1977) was thus an attempt to move beyond the biomedical model, though many argue that the biomedical model is still dominating the paradigm in healthcare, both in medicine and psychology (Engebretsen 2018; Engebretsen and Bjorbækmo 2019).

Fragmentation of Care. Although co- and multi-morbidity are the norm in the clinic, medicine and healthcare tend to be organised according to single diseases (Mercer et al. 2009; Parekh and Barton 2010; Vogt et al. 2014; Tómasdóttir et al. 2015). This specialisation of medical disciplines was brought about in order to enhance and deepen the specialists' skills and expertise. On the other hand, healthcare has been criticised for becoming increasingly compartmentalised, organised into distinct and sometimes isolated 'silos'. This means that patients with chronic and complex health complaints must go through the healthcare system by moving from specialisation to specialisation, treated as fragments, not as wholes (Kirkengen 2018).

Medicalisation of Life. In current healthcare there is the hope that a biomedical treatment, such as a drug, might ideally treat effectively even complex psychosocial phenomena (Ballard and Elston 2005). On the other hand, the expansion of the medical domain into most aspects of life, such as fertility, sexuality, sleeping patterns, angst, ageing and grief, has been criticised. Some are worried about placing experiences that all human beings are expected to have in the 'healthy-unhealthy' category. Ultimately, such a move suggests that it is imperative that we treat those experiences medically rather than accepting or living through them (Burgess 1993; Pilgrim and Bentall 1999; Moloney 2010).

1.2.2 Debating Methodology (Epistemology)

Information Gets Lost in Statistics. One ongoing debate is over what it means that clinical decisions should be evidence based. In evidence based medicine and practice, causally relevant evidence is taken to be statistical and population based, generated from large clinical studies. The aim is thus to ground the care of individuals in general knowledge about what is the most effective intervention in a studied population (Sackett et al. 2000). Critical voices have raised concerns about the tension between the public health interest in equality of care and the clinical needs of individuals. While evidence based policy is widely informed about what happens elsewhere, the worry is that causally relevant information about the unique local context is disregarded or lost (Cartwright and Hardie 2012).

The Importance of Mechanistic Knowledge. In evidence based medicine and practice, randomised controlled trials (RCTs) are seen as the gold standard for establishing causality (Howick 2011). This is because, thanks to their experimental design, a well conducted RCT is the best way to isolate one causal factor from potentially confounding factors and see whether it makes a statistical difference in outcome. In contrast, some argue that causal relationships cannot be established without the use of unquantifiable factors such as the theoretical knowledge coming from the laboratory and clinical science (Charlton and Miles 1998). This is parallel to the ongoing debate in philosophy of medicine on whether statistical knowledge must be accompanied by a theory of causal mechanism for the purpose of establishing causality (Russo and Williamson 2007; Osimani 2013; Anjum and Mumford 2018).

A Call for Phenomenology. For ethical reasons, it is not possible to establish whether psychosocial factors causally affect health in a negative way using clinical experiments. For instance, one cannot test the causal impact of childhood trauma, abuse, grief, psychological stress or social stigmatisation using RCTs, the gold standard for establishing causal relationships. One way to overcome this problem is to substitute RCTs with other statistical methods, such as cohort studies or case-control studies. This is still within the framework of evidence based medicine and practice. Other approaches emphasise instead individual uniqueness and phenomenology, urging the profession to change its focus to the whole patient experiencing the condition (Loughlin et al. 2018, see also Engebretsen, Chap. 11, Broom, Chap. 14 and Kirkengen, Chap. 15, this book).

Limited External Validity. In the health sciences, like in many natural sciences, causality is studied through experimentation, within controlled and somewhat artificial settings. Because of the need to control for confounders, clinical trials use strict inclusion and exclusion criteria for recruiting the participants. On one side, such controlled conditions increase the reliability of the experimental results, and the confidence that the observed result is actually due to the tested intervention. At the same time, however, this limits the external validity of the studies. When facing chronically ill patients, older patients, pregnant women, or even children, it is therefore not obvious that the results from clinical studies apply directly in respect to dosage, efficacy or even safety (Rothwell 2005, 2006; Baylis and MacQuarrie 2016).

1.2.3 Debating Practice

Upgrading Clinical Judgement and Knowledge. One motivation for the introduction of evidence based medicine and practice was to ensure that patients got the best available treatment, independently of the experience or preference of their healthcare practitioner. Rather than depending entirely on clinical judgement and the authority of expertise, treatment should be given according to the best scientific evidence, preferably from RCTs and meta-studies. Of concern among healthcare practitioners is how this depreciation of clinical judgement affects the clinical encounter. In particular, when practitioners are encouraged to use guidelines and computational tools to diagnose and make decisions about treatment, this leaves less room for their own clinical expertise and knowledge of the particular patient at hand. A worry is that, in the process of decision making, data from other patients will weigh more than the evidence from the person seeking care (Greenhalgh 2018).

Efficiency at the Cost of Individual Needs. New Public Management is an increasingly popular global phenomenon that started in the late 70s, with the aim of improving efficiency of public services by making them more similar to businesses (Diefenbach 2009). The introduction of New Public Management has affected the way in which healthcare is financed, organised and executed (Simonet 2008; Wyller et al. 2013). Health service delivery is supposed to be time- and cost-efficient and resources are allocated according to generic standards, such as type of diagnosis. On the other hand, proponents of person centered healthcare worry about the current trend towards package solutions and standardisation of care. This approach often hinders the assessment of individual needs, they warn. An alternative management ideology for the health services, according to these critical voices, could be one where the suffering individual, and not societal needs, has priority in setting the course of care. Calls for action have been raised among medical professionals, urging that the New Public Management approach is leading to a decay in healthcare, rather than to an improved quality and efficiency (Wyller et al. 2013).

1.3 Aims and Overview of the Book

This book is intended as an intellectual resource for clinicians and healthcare professionals who are interested in digging deeper into the philosophical foundations of their daily practice. It is a tool for understanding some of the philosophical motivations and rationality behind the way medicine and healthcare are studied, evaluated and practiced, at the interface of science and the humanities. We will show how a change in the ontological foundation could motivate a paradigmatic change in scientific methodology and clinical practice towards a genuine person centred approach, focusing on the whole person. In particular, this book illustrates the impact that our thinking about causality, both philosophically and conceptually, has on the clinical encounter.

By 'clinical encounter' we mean, in the broad sense, a consultation between the healthcare professional and the individual person seeking care. This is not limited to medical practice, but covers healthcare in general, including nursing, psychology, physical therapy, clinical pharmacy, rehabilitation, homecare services, as well as individual preventive care or any follow up of the patient. Although many of our examples come from medicine and general practice, the philosophical ideas should have a wider relevance also for these other parts of healthcare. If some of the philosophical biases that dominate current medical thinking actually limit the notion of evidence in a way that disadvantages the clinical encounter, then this needs to be critically discussed. Being able to identify the non-empirical foundation of scientific norms and practices is thus a first and necessary step for critically evaluating them.

In this book we want to prepare the ground for a genuine transdisciplinary discussion, not only between healthcare professionals and philosophers, but also one that expands to decision-makers and patients (Rocca et al. 2019). The main aim of this book is to engage and empower healthcare professionals to take part in changing and defining the premises for their own practice. After all, if clinical decisions should be based on evidence, this ought to be evidence that is relevant and well-suited for the clinic.

We have organised the book into two parts, *Philosophical Framework* and *Application to the Clinic*. The first part is written primarily from the philosophical perspective, by philosophers, and presents a singular overall framework. The second part is written primarily by clinicians who address some implications of the philosophical framework for different aspects of their own practice. The philosophical framework will thus be presented from different angles throughout the book, with more or less explicit focus on clinical practice. We hope that the diversity of voices, focus and perspective reflected in the chapters will contribute to make the philosophical ideas more accessible and relevant for practitioners with different professional backgrounds.

In the first part, we give an overview of the philosophical framework and themes of CauseHealth. In *Dispositions and the Unique Patient* we introduce the theory of dispositionalism and explain how this offers a better foundation for understanding causality in the individual case. In *Probability for the Clinical Encounter* we show how dispositionalism challenges the way we think about probabilities and propose an alternative – a singularist propensity theory – that we argue is better suited for the clinic. In *When a Cause Cannot be Found* we discuss how dispositionalism can throw some new light on medically unexplained symptoms, since this theory treats causal complexity, individual variations and medical uniqueness as essential features of causality rather than as problems for causality. Next, in *Complexity, Reductionism and the Biomedical Model*, we argue that a dispositionalist approach would deny any form of reductionism and render the biomedical understanding of health and illness deeply problematic. Finally, in *The Guidelines Challenge*, we discuss the tension between clinical guidelines, based on general medical knowledge and aimed toward standardisation, and their use in the clinical encounter, based on local knowledge about the patient and aimed toward tailored interventions.

In Part II, eight practitioners and one patient from the CauseHealth network translate the philosophical ideas into a clinical setting. In their contributions, they show how philosophical reflections concerning foundational issues have influenced their own practice and how they understand and deal with health and illness. This part has nine chapters. *The Complexity of Persistent Pain – A Patient's Perspective* is written by Christine Price who suffers from chronic pain. Price explains how she uses the dispositionalist framework to understand, model and manage her own chronic pain after she learned about this from her physiotherapist Matthew Low. Low is the author of *Above and Beyond Statistical Evidence. Why Stories Matter for Clinical Decisions and Shared Decision Making*. In this chapter he explains why patient narratives and dispositionalism are valuable resources for shared clinical decision making. In *Causality and Dispositionality in Medical Practice*, general physician and clinical pharmacologist Ivor Ralph Edwards discusses the tension between the need of a full phenomenological, dispositional and causal evaluation for making better diagnoses and the practical restrictions of the clinical reality.

Following up on these challenges, *Lessons on Causality from a Clinic for Patients with Severe Obesity* is written by senior consultant and general practitioner Kai Brynjar Hagen. Hagen describes how conversations dedicated to the first encounter with the patients allowed him to get closer to a causal diagnosis of their obesity, suggesting a causal therapy rather than a purely symptomatic one. Next, in *Reflections on the Clinician's Role in the Clinical Encounter*, psychotherapist Karin Mohn Engebretsen illustrates how the clinician's own personal and professional background influences the therapy in positive or negative ways and explains why clinicians ought to be conscious about what they bring with them to the clinical encounter. In *The Relevance of Dispositionalism for Psychotherapy and Psychotherapy Research*, psychologist Tobias Gustum Lindstad explores the influence of the evidence based framework on psychotherapy and proposes a dispositional revival of the profession. *Causal Dispositionalism and Evidence Based Healthcare*, written by physiotherapist Roger Kerry, discusses whether a new evidence based practice framework can be offered, one underpinned by a dispositional ontology, and reflects on how this would look. Next, *The Practice of Whole Person-Centred Healthcare*, by immunologist and psychotherapist Brian Broom, is an account of the emergence in New Zealand of a non-dualistic, whole person centred form of clinical practice, particularly in relation to the treatment of physical disease of all kinds. In *A Broken Child – A Diseased Woman*, general practitioner Anna Luise Kirkengen contrasts the personal biography of a patient, which is a history of violation, to her sickness histories as documented in the medical records. The chapter is a powerful reminder of why medicine and healthcare must be genuinely person centred in order to obtain crucial information for understanding, diagnosing and treating patients.

In its totality, this book reinterprets what a genuine person centered approach should entail; from ontological foundation to norms of methodology and practice. This means that even those already educated within a person centered framework might have some of their preconceptions challenged by the dispositionalist theory presented here. We conclude the book by proposing a list of CauseHealth recommendations for a paradigmatic change in medicine and healthcare.

References and Further Readings

Anjum RL (2016) Evidence-based or person-centered. An ontological debate. Eur J Pers Cent Healthc 4:421–429
Anjum RL, Mumford S (2018) Causation in science and the methods of scientific discovery. Oxford University Press, Oxford
Andersen F, Anjum RL, Rocca E (2019) Philosophical BIAS is the one bias that science cannot avoid. eLife 8:e44929. https://doi.org/10.7554/eLife.44929
Arnaudo E (2017) Pain and dualism: which dualism? J Eval Clin Pract 23:1081–1086
Ballard K, Elston MA (2005) Medicalisation: a multi-dimensional concept. Soc Theory Health 3:228–241
Baylis F, MacQuarrie R (2016) Why physicians and women should want pregnant women included in clinical trials. In: Baylis F, Ballantyne A (eds) Clinical research involving pregnant women. Springer, Cham
Burgess M (1993) The medicalization of dying. J Med Philos 18:269–280
Cartwright N, Hardie J (2012) Evidence-based policy. A guide to do it better. Oxford University Press, Oxford
Charlton BG, Miles A (1998) The rise and fall of EBM. Q J Med 91:371–374
Diefenbach T (2009) New public management in public sector organizations: the dark sides of managerialistic 'enlightenment'. Pub Adm 87:892–909
Engebretsen KM (2018) Suffering without a medical diagnosis. A critical view on the biomedical attitudes towards persons suffering from burnout and the implications for medical care. J Eval Clin Pract 24:1150–1157
Engebretsen KM, Bjorbækmo WS (2019) Burned out or "just" depressed? An existential phenomenological exploration of burnout. J Eval Clin Pract. https://doi.org/10.1111/jep.13288
Engel GM (1977) The need for a new biomedical model: a challenge for biomedicine. Science 196:129–136
Eriksen TE, Kerry R, Lie SAN et al (2013) At the borders of medical reasoning – aetiological and ontological challenges of medically unexplained symptoms. Philos Ethics Humanit Med 8:1747–1753
Greenhalgh T (2018) Of lamp posts, keys, and fabled drunkards: a perspectival tale of 4 guidelines. J Eval Clin Pract 24:1132–1138
Howick J (2011) The philosophy of evidence-based medicine. BMJ Books/Wiley-Blackwell, Oxford
Kirkengen AL (2018) From wholes to fragments to wholes—what gets lost in translation? J Eval Clin Pract 24:1145–1149
Kuhn T (1962) The structure of scientific revolutions. University of Chicago Press, Chicago
Loughlin M, Mercuri M, Parvan A et al (2018) Treating real people: science and humanity. J Eval Clin Pract 24:919–929
Mercer SW, Smith SM, Wyke S et al (2009) Multimorbidity in primary care: developing the research agenda. Fam Pract 26:79–80

Moloney P (2010) 'How can a chord be weird if it expresses your soul?' Some critical reflections on the diagnosis of Aspergers syndrome. Disabil Soc 25:135–148

Osimani B (2013) Until RCT proven? On the asymmetry of evidence requirements for risk assessment. J Eval Clin Pract 19:454–462

Parekh AK, Barton MB (2010) The challenge of multiple comorbidity for the US health care system. J Am Med Assoc 303:1303–1304

Pilgrim D, Bentall R (1999) The medicalisation of misery: a critical realist analysis of the concept of depression. J Ment Health 8:261–274

Rocca E, Copeland S, Edwards IR (2019) Pharmacovigilance as scientific discovery: an argument for trans-disciplinarity. Drug Saf. https://doi.org/10.1007/s40264-019-00826-1

Rothwell PM (2005) External validity of randomised controlled trials: "to whom do the results of this trial apply?". Lancet 365:82–93

Rothwell PM (2006) Factors that can affect the external validity of randomised controlled trials. PLoS Clin Trials. https://doi.org/10.1371/journal.pctr.0010009

Russo F, Williamson J (2007) Interpreting causality in the health sciences. Int Stud Philos Sci 21:157–170

Sackett DL, Straus SE, Richardson WS et al (2000) Evidence-based medicine: how to practice and teach EBM, 2nd edn. Churchill Livingstone, Edinburgh

Simonet D (2008) The new public management theory and European health-care reforms. Can Public Adm 51:617–635

Tómasdóttir MO, Sigurdsson JA, Petursson H et al (2015) Self reported childhood difficulties, adult multimorbidity and allostatic load. A cross-sectional analysis of the Norwegian HUNT study. PLoS One 10:e0130591

Tyreman S (2018) Evidence, alternative facts and narrative: a personal reflection on person centred care and the role of stories in healthcare. Int J Osteopath Med 28:1–3

Vogt H, Ulvestad E, Eriksen TE et al (2014) Getting personal: can systems medicine integrate scientific and humanistic conceptions of the patient? J Eval Clin Pract 20:942–952

Wade DT, Halligan PW (2004) Do biomedical models of illness make for good healthcare systems? BMJ 329:1398

Wyller VB, Gisvold SE, Hagen E et al (2013) Reclaim the profession! Tidsskr Nor Legeforen 133:655–659

Open Access This chapter is licensed under the terms of the Creative Commons Attribution 4.0 International License (http://creativecommons.org/licenses/by/4.0/), which permits use, sharing, adaptation, distribution and reproduction in any medium or format, as long as you give appropriate credit to the original author(s) and the source, provide a link to the Creative Commons license and indicate if changes were made.

The images or other third party material in this chapter are included in the chapter's Creative Commons license, unless indicated otherwise in a credit line to the material. If material is not included in the chapter's Creative Commons license and your intended use is not permitted by statutory regulation or exceeds the permitted use, you will need to obtain permission directly from the copyright holder.

Chapter 2
Dispositions and the Unique Patient

Rani Lill Anjum

> I have been working as a regular GP for many years and experienced how useful it is to know patients as persons. Through repeated encounters over time, I became familiar with many of my patients as persons—who they were and how they lived their lives—whether I was aiming for it or not.
>
> Although such information may be of medical relevance, it is rarely written down in the medical records. In many aspects, it is tacit knowledge. Nevertheless, as General Practitioners we use this kind of knowledge all the time in everyday medical practice, tailoring the approach and follow-up of individual patients, especially when we are dealing with the patients we see the most.
>
> There is also a growing body of research documenting that adverse life experiences can have a decisive impact on people's health. However, traditional biomedicine, the dominant perspective through which today's medical practice is conceptualised, has placed little emphasis on expert findings, such as those within modern stress physiology, indicating that biographical and person-related knowledge has medical relevance.
>
> Bente Prytz Mjølstad, 'Does your regular GP know you – as a person?', CauseHealth blog (https://causehealthblog.wordpress.com/2017/11/09)

2.1 The Similar and the Unique

From the biomedical perspective, medicine primarily deals with what is common for all: cells, tissues, organs, anatomy and biological processes and interactions. In the clinical encounter, however, one has to also deal with what is particular and unique. But how much space should the practitioner give to evidence that is specific to the single patient? Given that time and resources are limited, the highest priority must be given to the clearly defined medical facts. Then, *if* there is time, one can talk to the patient about other and more personal matters. Or so one might argue.

R. L. Anjum (✉)
NMBU Centre for Applied Philosophy of Science, Norwegian University of Life Sciences, Ås, Norway
e-mail: rani.anjum@nmbu.no

© The Author(s) 2020
R. L. Anjum et al. (eds.), *Rethinking Causality, Complexity and Evidence for the Unique Patient*, https://doi.org/10.1007/978-3-030-41239-5_2

In the quotation above, general practitioner Bente Prytz Mjølstad offers a different perspective. She suggests that knowing the patient as a person might also help the clinician to better understand their medical condition and medical needs. Perhaps, then, knowledge of what is unique to a patient ought to be given a higher priority in the clinic. Or would that take time and resources away from what is most important: to understand, diagnose and treat the patient? In the CauseHealth project, we have met a number of practitioners who emphasise the importance of patient narratives, and who use the patient's perspective and story as a source of valuable medical information (many of these have contributed to Part II of this book). Immunologist and psychotherapist Brian Broom describes this as follows:

> Sometimes people wonder why I am so keen on the 'story'. It is simply that we clinicians who want to practice in a whole person way need practical doorways into the world of the whole person, and especially that part of the person's world not accessed by the normal biomedical approach to disease. The latter, as currently practised, focuses on the expertise of information-holding, the power of drugs and physical interventions, activities directed at restoring, repairing and compensating for 'mechanical' deficits and distortions, and so on. I greatly value the benefits of much of this.
>
> But listening for the patient's story opens up an entirely different world, and its power derives from quite a different set of assumptions, attitudes and relational values. Asking for a story may seem a simple matter but the implications are hugely important. In reality, most of the stories implicated in illness are relational stories: of trauma, failure, loss, abuse, abandonment, manipulation, and much more. We are all fundamentally relational…
>
> We don't need to be a psychologist or psychotherapist to start this process, or to make simple connections, or to give the patient permission and encouragement to consider the connections. The interaction doesn't have to be perfect, or prolonged, or all done at once. Patients know that we are persons too, and have limits on our capacities. We can be good-enough.
>
> Brian Broom, 'Imagination and its companions', CauseHealth blog (https://causehealth-blog.wordpress.com/2017/07/03)

Clinicians might see an advantage in knowing more about their patients' stories, for the purpose of finding the causes of health and illness and for making predictions about treatment and recovery. But there is little support for patient stories within current medical thinking. In evidence based medicine and practice, information from the single patient is not generally treated as strong evidence, at least not of causality. One might even refer to the experience of individuals as 'subjective' or 'anecdotal', suggesting that their stories are relevant mainly for themselves and not scientifically valid for claiming, for instance, that a certain factor contributed to a certain condition. A question we should ask is then: *at what point does information from individual patients become causal evidence* (see also Kerry, Chap. 13, this book)? Is there a threshold at which anecdotes transform into evidence, for instance when there is a sufficient number of individuals who report similar experiences?

Say a patient reports experiencing a possible side effect from a prescribed medicine. If no one else using the medicine has reported the same side effect, one might be reluctant to conclude that the medicine caused that effect, and for good reasons.

But if, after a few years on the market, a sufficient number of people using the medicine report the same reaction, it might be concluded that it is a side effect after all.

From an *epistemological* perspective, meaning from the perspective of what we can or cannot know, this is reasonable. If one knows little about the medicine's causal mechanisms, a single report is not itself sufficient evidence of causality. But if one thinks that 10, 100 or 1000 reports amounts to sufficient causal evidence, then didn't causality happen within each of these individuals? So even if the first patient reporting the effect was not sufficient evidence of this causal link, it does not follow that causality did not happen also in that instant. Lack of evidence does not imply lack of causality: we cannot conclude that there is no causality happening in that particular person, ontologically, simply because we, epistemologically, lack evidence of causality. Ontologically, meaning from the perspective of what does or does not exist, if causality happens, it does so in the particular instance and does not depend on what happens in other cases to similar patients.

> Simply put...
>
> Ontology concerns reality: what exists or happens in the world.
>
> Epistemology concerns knowledge: what we can and cannot know about reality.
>
> Example: Ontologically, one might have a rare genetic disease even if one has yet not discovered it. Lack of knowledge does, therefore, not entail lack of reality. Ontologically, the disease exists, but epistemologically, we don't have any evidence or symptoms of it. In other situations, there might be symptoms or evidence of some disease that is not actually there, such as when a cancer screening gives a false positive. Epistemologically, there was evidence of cancer, but ontologically, there was no cancer.

Why is the distinction between epistemology and ontology important? First of all, if causality happens in the particular case (ontologically), this means that each individual patient represents a valuable source of evidence (epistemologically), also for causality. Secondly, this distinction also points to a tension in how we understand causality as a phenomenon. Philosophically it boils down to the question of whether there in principle could be a case of a unique causal event. *Could causality happen only once and never be repeated throughout the whole history of time?* This is an ontological question and not the same as the epistemological question of whether we could ever scientifically establish causality for such a unique case.

This question is relevant for the clinical encounter because it matters in terms of how one deals with medical uniqueness and individual variation. Should we place much weight on single case reports or patient stories of seemingly unique effects or

Fig. 2.1 The original evidence hierarchy of EBM

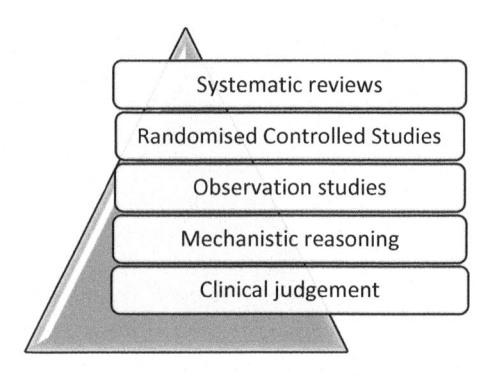

- Systematic reviews
- Randomised Controlled Studies
- Observation studies
- Mechanistic reasoning
- Clinical judgement

experiences? Or must we wait for results from larger clinical studies to even consider causality?

In the original evidence hierarchy of evidence based medicine (EBM) (Fig. 2.1), patient narratives, or even case reports, do not count for much, at least not with respect to causality. The idea is that unless there is repetition, and preferably many repetitions, one simply cannot know if something is causally related. This places EBM within the philosophical tradition of empiricism and its emphasis on the observable. We will now see how empiricism has influenced our understanding of causality.

2.2 Empiricism: Causality Requires Repeated Observations

Historically, empiricism is the most influential view for how we understand causality, in philosophy and science in general. This view has itself been largely shaped by the empiricist philosopher David Hume and his famous work *A Treatise of Human Nature* from 1739. Hume was critical of ontology: he was skeptical about making claims about what ultimately exists. Indeed, he believed that all we can really know about the world is restricted by what we can experience through our senses. This is the empiricist assumption, namely that our observations are the only evidence we can have of reality. Positivism is the idea that science should be strictly empirical, and only deal with what can be observed and measured.

This scepticism toward ontology also meant that Hume was critical of any theory that says more than what can be backed up by data. If we try to explain what we observe by appealing to some general, underlying principle, this would be to say more than what we have evidence of. In practice, however, this might mean that all scientific theories have a speculative element to them, unless they simply report the available data. For Hume, therefore, as for any empiricist, epistemology trumps ontology.

> Simply put…
>
> Empiricism is the idea that we can only know what can be experienced through our senses. This means that all scientific knowledge should ultimately come from observation data.
>
> Positivism is a strict empiricist view of science, stating that science should only deal with what can be established through observation and measurement, and that everything else is metaphysical speculations.

When analysing the concept of causality, Hume used the example of the billiard ball table. Here, we think that hitting the object ball with the cue ball causes the object ball to roll. But what do we actually observe on the table? Hume found three observable features of causality:

1. The cause is perfectly correlated with its effect, so that every time the cause happens, the effect follows. He called this the *constant conjunction* of cause and effect.
2. The cause always happens before the effect. He called this *temporal priority*.
3. There must be some contact between cause and effect in time and space, so that the cause and effect meet. He called this *contiguity*.

What Hume could not observe, however, was any form of link or necessary connection between cause and effect. We cannot observe that also the next time the cause occurs, the effect will necessarily happen. If we have not observed this yet, we cannot infer that it will happen, even if we are convinced that it will.

To assume that what we have observed until now will be what we observe also in the future, is what Hume called an inductive inference. These inferences are logically invalid, since the conclusion goes beyond the premises. So we have seen something happening in the past, but then we infer that it will be the same in the future. This is something that we cannot know yet. When it comes to causality, therefore, all we can know is that it is a form of regularity, as specified by 1–3 above. We cannot infer from this that there is a causal law that guarantees the same pattern of regularity in the future, as this is yet to be observed.

> Simply put…
>
> Inductive inference is the process of drawing a conclusion by going beyond the available evidence. For example, one could infer a conclusion for a whole population from the results of a study conducted on a representative sample of such a population. Since the conclusion goes beyond the premises, we cannot be sure that inductive inferences are logically valid.

Hume's is a very strict form of empiricism. To follow it up scientifically, we would have to constrain ourselves from making any form of general or theoretical claims that also involve future events. The law of gravitational attraction, for instance, might have worked in the past and the present, but whether it will work tomorrow is not yet evident. A causal law is then nothing but a claim that the cause and the effect have been repeatedly observed to follow one another in a certain way in the past.

This is now known as the *regularity theory of causality*, in which repetition is the key to calling something causality. From this perspective, the same cause should always give the same effect, at least under the same or similar conditions. Hume was quite insistent on this criterion for causality: 'The same cause always produces the same effect, and the same effect never arises but from the same cause' (Hume 1739, I, iii: 173). In the clinic, we might refer to a certain sub-population of patients that share some important features, such as their diagnosis. If these patients are given the same treatment, we might expect that they should also get the same effect from it. But we know that this is not generally the case. How can this be explained philosophically?

There is an important consequence of Hume's principle, that *same cause, gives same effect*. If the effect is different, it can only mean that something was different in the cause, or in the background conditions. Hume seems to think of this as an undeniable truth about causality:

> The difference in effects of two resembling objects must proceed from that particular, in which they differ. For as like causes always produce like effects, when in any instance we find our expectation to be disappointed, we must conclude that this irregularity proceeds from some difference in the causes. (Hume 1739, I, iii: 174)

Hume would then say that if two patients with the same diagnosis get different effects from the same treatment, there must be some other difference between them that is causally relevant. The question is which individual differences are causally relevant for the effect and which are not. We might expect sex, age and weight to be relevant, but many treatments work across these differences. One might still expect that if everything were the same between two patients, the same intervention should produce the same outcome. If so, this would be in accordance with Hume's understanding of causality.

Another consequence of Hume's theory is that there can be no unique instance of causality. Without the possibility of repetition, one cannot check whether the same cause always produces the same effect. Hume accepts this and even goes as far as to say that if the creation of the universe happened only once, it could not count as causal. The alternative to this view is the position called 'causal singularism', which will be discussed in the next section.

Although many philosophers might disagree with Hume in one or more of these features of causality, they might still agree with his empiricist starting point: (i) that the causal link itself is not directly observable and (ii) that causality must therefore be inferred from what we can actually observe. In science, however, there seems to be a general acceptance of Hume's idea that same cause gives the same effect, and that any difference in the effect must come from a difference in the cause.

How does this relate to how we understand causality in clinical research? Assuming the empiricist ideal of science, causal relations should then be established from observation data, for instance of an intervention and its outcomes. From such data, one might be able to observe regularities, difference-makers or probability-raisers, all of which can be detected via statistical methods. If causality is accessible via observation data in this way, causal hypotheses and theories could in principle be generated directly from data.

Crucial for this understanding of causality is that one needs repetition. Statistical methods require large samples, or at least more than one or two individuals. In cases of individual variation, one at least needs an actual or assumed sub-population of which this patient is thought to be representative. A problem with a view of causality based on repetition, is that it leaves no room for causal uniqueness, as will be explained further in the next section. If all patients are different, having a unique biology, biography, life-style, diet, and so on, then no sub-group will perfectly represent them.

If repetition is a requirement for establishing causality, all causally relevant differences between individuals seem to fall outside the scope of a single study. From the clinical perspective this is a practical problem. *How to deal with causality in cases of large individual variation?* Or perhaps we should say in case of uniqueness *when $N = 1$?* This question was one of the main drivers for the Cause Health project.

2.3 Dispositionalism: Causality Happens in the Single Case

There are two ways to think about causal uniqueness. One is to think of it as a *problem* of causality, because we are then lacking the possibility of confirmation from other similar cases. The other way is to think of causal uniqueness as *typical* of causality, and therefore as the default expectation in any causal assessment. Causal dispositionalism represents the latter view, called 'causal singularism'. This section presents a brief overview of the dispositionalist theory of causality and explain why it gives ontological and epistemological weight to the particular over the general.

The theory of causal dispositionalism was first introduced in Mumford and Anjum (2010) and is described in detail in their book *Getting Causes from Powers* (Mumford and Anjum 2011). We will now explain why the individual patient and their narrative should be at the heart of causal matters from a dispositionalist perspective.

> Simply put…
>
> Causal singularism is the ontological view that causality happens in the particular case and does not require repetition.
>
> Example: The first person to die from a rare disease is a single and unique case up until the next person dies from it. But even in the first person, the disease caused their death, ontologically. So even if one might need more cases to establish causality, epistemologically, either in animal models or in a clinical study, causality happens in each individual case.

2.3.1 Causes Are Dispositions

Dispositions are also referred to as 'causal powers', 'abilities' (Mumford 1998) or 'capacities' (Cartwright 1989). They refer to what something can do. A sharp knife can cut, a medication can heal, and a virus can make an otherwise healthy person ill. A disposition is a type of property, but one that can exist unmanifested. Examples of dispositional properties are fragility, flammability, toxicity and fertility. A substance is toxic even when it is not harming anyone. And a person can be fertile throughout their life without reproducing. Causality typically happens when dispositions manifest themselves. A fertile woman becomes pregnant, toxic arsenic kills or some explosive substance explodes. In this sense, the dispositional property is a cause and the manifestation is an effect.

Whether something or someone has a disposition is not always observable until it is manifested. The 'proof' of a disposition's existence thus lies in its manifestation. Empiricist philosophers have asked how we can even know that dispositions are real if they cannot be observed. Some dispositions might be possible to tease out by an intervention or a test, such as a fertility test. But there will always be dispositions that we simply cannot know of until they are manifested, and perhaps not even then. A person might have early stage cancer without manifesting any observable symptoms, but the causal process has nevertheless started. A disposition is thus not a pure possibility, like the possibility of flying cars in the future. It is a potentiality that exits in the world here and now as a real possibility in the properties of things.

Since empiricists trust only what can be observed (*observability* being another disposition), they take manifestations to be real but see dispositions as merely possibly real. This seems a plausible conclusion if we think that the dispositions are nowhere until they are manifested. But many dispositions are clearly present also before they manifest. For instance, a sharp knife has a disposition to cut through skin in virtue of the shape and material of the knife. If a knife was made of a material that was too soft to penetrate the skin, it would lack that disposition.

According to Hume, our knowledge about dispositions is inferred from what we have already observed elsewhere. Hume and neo-Humean philosophers, such as David Lewis, Stathis Psillos and Helen Beebee, are therefore sceptical of dispositions. They would therefore not include dispositions in their ontology unless they are analysed into something observable (Mumford 1998, 2004). The only reason why we think a wine glass is fragile, one might say, is because we have seen other wine glasses break from very little impact elsewhere. Whether a particular wine glass is actually fragile is thus something we cannot know until it breaks.

Epistemologically, this might be the case for many dispositions. But ontologically, at least, once the glass actually breaks, doesn't this mean that it was in fact fragile? If we had to wait until a manifestation occurred in every case before we believed in its dispositions, we could not say that a nuclear power plant was explosive unless it explodes. Dispositionalists would therefore reject the strict empiricist principle, and argue that some things could be real even if they are not directly observable.

Dispositions are seen as plausibly real because they can explain what actually happens – the underlying principles of the behaviour of things. Causal effects without underlying dispositions would on this view be unaccounted for and remain an ontological mystery. The Humeans, on the other hand, would rather see everything that is not observable as representing the real mystery, but their motivation for saying so is primarily epistemological: we could not possibly know of something's existence (ontology) unless we can observe it (epistemology).

Although some philosophers are sceptical of dispositions, these properties seem to play an important role in our lives. That we take dispositions seriously can be seen in how they influence our behaviour. We are careful around explosive, flammable or poisonous substances, and we don't expose ourselves unecessarily to contagious diseases or let our children play with sharp knives. As Stephen Mumford puts it in the opening section of his book *Dispositions* (1998: 1), referring to Nelson Goodman, this is a world of threats and promises. And our behaviour very much reveals our understanding of dispositions as real and important.

How is this relevant for the clinic? One might from the observation of a heavy smoker's lungs see that they are disposed to emphysema, chronic bronchitis and lung cancer. And if a person's arteries are clogged by arterial plaques, aren't they disposed to reduced blood flow and therefore to heart attack and stroke? In this sense, the current situation points toward a possible, or even likely, future. The dispositions might reveal the direction toward which the situation is heading: what tends to be (Anjum and Mumford 2018a). Dispositions are thus useful for making prognoses for illness and recovery, but also for making the correct diagnosis. Since many symptoms could be the manifestations of a range of illnesses, it is important for choosing the right treatment that it targets the right disposition. If a headache is caused by stress, the treatment will be different from if the headache is caused by a tumour in the brain. Similarly, if chronic depression is caused by a biological disposition or by childhood trauma or abuse, or both, the treatment scheme should reflect this difference (see also Hagen, Chap. 10, this book).

2.3.2 Causes Are Intrinsic

Dispositions are typically intrinsic properties, belonging to some particular individual or entity. That dispositions are intrinsic is crucial for the purpose of causality, since we should not say that a drug works unless it has an intrinsic property to bring about its effect. This is why medical interventions are typically tested against a placebo, to check whether the effect on recovery comes from the intervention or from the patient's own expectations of recovery. We might say that the placebo effect is a manifestation of the patient's dispositions, and not of dispositions belonging to the intervention.

That dispositions are intrinsic does not mean that they have to belong to an individual. Many dispositions belong to entities that are higher-level or even abstract. A community can be supportive, friendly, homophobic or xenophobic, for instance. A family can be loving or dysfunctional. A work environment can be stimulating or draining and a legal system can be racist. Some dispositions might only emerge at group level. As a community of people, we have social dispositions related to communication, relationships, politics and law. Arguably, none of these are intrinsic to the individual but emerge as a result of interactions with others (Anjum and Mumford 2017, see also Rocca and Anjum, Chap. 5, this book).

> Simply put...
>
> A disposition is an intrinsic property that can exist unmanifested and gives its bearer a causal power, ability or capacity.
>
> Example: Someone can have a disposition toward a disease that is never manifested, just like a glass can be fragile without being broken.

2.3.3 Causality Is Complex

We said that whenever dispositions are manifested, causality happens. A fertile woman becomes pregnant, a fragile glass breaks, a medicine cures an illness: these are all examples of causality. It is, however, important to recognise that all such manifestations are a result of multiple dispositions working together. It takes much more than a fertile woman and her eggs to become pregnant. Without the sperm from a fertile man, for instance, and a prepared uterus with the correct balance of hormones, the pregnancy will not happen. All these are what we call the *manifestation partners* for pregnancy, a term initially used by Martin (2008). That something is an appropriate manifestation partner for a disposition means that they can produce an effect together that neither of them could have produced on their own.

From a dispositionalist perspective, all causality is complex in this sense, requiring the interaction of one or more mutual manifestation partners. When a match is struck and lights, this effect is caused by the striking as well as the flammable match, the dry wood and the oxygen. But rather than treating one of these as *the* cause and others as background conditions, they are all causes of the effect in virtue of their own dispositions. Some of these dispositions might be necessary for the effect, while others might be thought of as triggers. Still, everything that contributes to producing the effect are on the dispositionalist view *causes*.

Causal complexity is particularly important to recognise in the clinic, since one cannot focus only on the medical intervention when treating a patient. What an intervention does on population level is one thing, but in this case, it will be interacting with a particular individual. Unless this individual is an appropriate manifestation partner for the intervention, it will not be able to do its causal work. For example, antibiotics tend to treat infections, but only in interaction with a non-resistant bacterium and proper conditions for being absorbed by the patient's body. Since most medical interventions have more than one disposition, a patient might be a non-responder to the targeted effect, but still a mutual manifestation partner for one or more of the non-targeted effects (see Edwards, Chap. 9, this book).

Simply put…

Mutual manifestation partners are a pair or set of dispositions that can do causal work in interaction with each other that they cannot do on their own.

Example: When a match is struck and lights, this is the manifestation of the flammable tip, the dry wood, the presence of oxygen, and so on.

When choosing a treatment for a particular patient, therefore, one should try to learn more about the dispositions of the patient who will be interacting with the treatment, as well as looking into the dispositions of the treatment. Most of the causally relevant dispositions in a treatment situation will actually come from the patient and their unique causal set-up, including medical history, genetics, diet, life situation and biography. This is why rich patient evidence is important, but it is also why it is important to understand causal mechanisms. Such mechanisms will tell us how the treatment works in the body, but also how the various dispositions of the patient might interact with the treatment.

How important is it for the clinic to have mechanistic evidence? Evidence based medicine and practice emphasise statistical evidence over mechanistic knowledge for establishing whether an intervention works. One argument for this is that our knowledge about pathophysiological mechanisms is at any time incomplete, and might be wrong. Because of this, some EBM proponents argue (Howick 2011), prediction about the effectiveness of an intervention should be based on population trials rather than on fallible mechanistic thinking. However, for the purpose of finding out how an intervention works, knowledge of causal mechanisms seems necessary. Russo and Williamson (2007) proposed what has been called the Russo-Williamson thesis. This thesis says that in order to establish causality, one needs both statistical evidence and evidence of mechanisms. Indeed, the correlations that are yielded from population studies are not necessarily causal.

Let's take an example. Use of paracetamol is correlated to a higher incidence of asthma, but this association could be due to confounding by indication. This means that the onset of asthma could be due to frequent infections rather than to the concomitant use of paracetamol. In order to establish whether the correlation is causal or not, it is necessary to understand the mechanism by which the cause brings about the effect.

Russo and Williamson (2007) also argue that mechanistic knowledge is needed to plan the experimental design of clinical trials, as well as for the interpretation of the results from such studies. Rocca (2018) adds to these that knowledge of the causal mechanisms underlying medical phenomena is necessary to evaluate complex evidence, and to judge which population study we need to trust when different studies give conflicting results. Gillies (2018) argues that mechanistic knowledge is needed, not only to establish causal hypotheses about the cause of illness, but also to develop an appropriate treatment and for evaluating the safety of a treatment.

From a dispositionalist point of view, knowledge of causal mechanisms is crucial for understanding causality. On this perspective, the mechanism is a complex and contextual matter, and includes the types of dispositions that are involved in the causal process, how they interact, and also a potential for dispositions interfering with and altering the causal process.

Without any understanding of causal mechanisms, it is difficult to predict how a treatment will affect a particular patient and their unique set of dispositions. Statistical data from other patients might help reveal causal mechanisms, but there will always be causally relevant differences between individual patients that influence the outcome of an intervention. The more knowledge we have of the dispositions of an intervention (both beneficial and harmful) and of the patient (including vulnerability), the better our predictions will be for how the treatment will work for that patient. This is also why one always needs to know which other medications a patient is taking. While two of the drugs might be appropriate mutual manifestation partners for the desired outcome, there might be other drugs that interfere and alter that outcome. This brings us to the context-sensitivity of causality.

2.3.4 Causality Is Context-Sensitive

A cause will tend to produce different effects in different contexts, depending on what else it interacts with. We have seen that this is a problem for empiricism: in order to define C as the cause of a certain outcome, we need to observe the same outcome every time we have C. For Hume, if the outcome observed is different, then we cannot talk about the same cause. This is why it is hard to define causes outside of experimental isolation. However, the story changes if we think of a cause as an intrinsic disposition that might exist without being manifested, something we saw was unacceptable for Hume. The same disposition might tend to produce different effects. In fact, anything else should be surprising. How?

Because effects are produced, not by single dispositions but multiple, we cannot expect that the same causal intervention will always produce the same effect. Different contexts will give different effects, and we should not expect that two contexts are ever exactly the same. This is essential in the clinic. Assuming that all patients are different in at least some dispositions, each patient will represent a unique set-up of mutual manifestation partners for a treatment. One patient will therefore be a different mutual manifestation partner for the treatment than another patient. So even if the treatment works in the intended way in both patients, meaning that the treatment has the same disposition in both, the two patients might get different side effects, or the treatment might work with different strength or momentum in each of them. The outcome, or manifestation, of a disposition can thus be different in different contexts, but this does not mean that the disposition of the intervention (e.g. the drug) was different. This is why we urge that the same cause does not always give the same effect. By thinking this way we need some additional strategies for making claims about the disposition of the drug, other than Hume's perfect regularity of cause and effect. For instance, we might know the drug's disposition because we know how it works.

We can illustrate this point with an example of extreme context-sensitivity, where the same causal disposition tends to produce widely different effects in different contexts. Antiarrythmic drugs have the disposition to calm irregular heartbeat, by altering the electrophysiology of myocardial cells at a faster heart rate. Such alteration is obtained by inhibiting the fast sodium channels. This in turn prolongs the action potential refractory period in some of the myocardial tissues. This disposition is an intrinsic property of the antiarrythmic drug, and since we are aware of this property and how it works, we can say it is present also when it is not manifested. In some patients, the same types of drugs can on the contrary worsen irregular heartbeat, by inhibiting a different type of electric flow through the cell membrane, for instance by blocking potassium channels. In this case, the manifestation is different because of a different set up of mutual manifestation partners. However, we can still say that the disposition of the drug to block the sodium channel was present and the same. What is different is the way in which the disposition of the drug manifested itself in patients with different sets of dispositions. This is what it means to say that causality is context-sensitive from a dispositionalist approach.

Could we plausibly say that the same intervention amounts to the same treatment in two different patients, if one patient's symptoms are alleviated while the other patient's symptoms are aggravated? In the first case, the intervention produces the effect it was intended to produce. But in the second case, another effect happened. Was it a side effect of the drug? Or did the patient interfere with the drug, preventing the effect from happening? In this case, it seems more accurate to say that the two causal set-ups produced different effects because of the different mutual manifestation partners involved.

2.3.5 Modelling Causality

We can model causality in the single case using the vector model (Fig. 2.2). The model was developed in Mumford and Anjum (2011), adapted to the clinical scenario in Low (2017, see also Low, Chap. 8, this book) and later used by Price (see Price, Chap. 7, this book) to understand and manage the complexity of her chronic condition. In the vector model, the current situation is represented by a vertical line, on a quality space between two outcomes, F and G. In a patient with a chronic condition, such as irritable bowel syndrome, F might represent lack of gastrointestinal symptoms while G might represent continuous symptoms. Then we add the dispositions which are in place simultaneously and contribute to either of these outcomes as vectors. Vectors allow us to model two important features of dispositions: their degree of strength and their direction. Say, for instance, that for this patient consumption of fatty meals disposes toward gastrointestinal symptoms to a greater degree than salt or sugar. In that case, this should be reflected in the length of the vectors. One should not only include the dispositions that dispose toward the appearance of symptoms, of course, but also those that dispose away from them. The patient might experience less symptoms when he exercises regularly for instance, or after a good night of sleep. The resultant vector R thus shows whether the overall tendency disposes toward F or G, and how much.

An important feature of the vector model is the qualitative nature of the vectors. The length of the vector is not reflecting a numeric or statistical tendency, of how often the effect happens in a particular population. This means that the length of the vectors, as well as the type of disposition represented by the vectors, will vary from one individual to another. The reason for this is that the vector model should represent the singularism of causal dispositionalism: that causality happens in the unique particular context. While one person will tend to an impoverished gastrointestinal flora because of intensive pharmacological treatment, another person might not share this disposition, while still being overall disposed to irritable bowel syndrome. The length and direction of the vector should therefore be based on what is the case for a particular person at a particular time. One thing is that general scientific knowledge (generated by a plurality of evidence, including population studies) can be useful to *suggest* which causal disposition might be at play in the case of this

Fig. 2.2 The vector model of causality

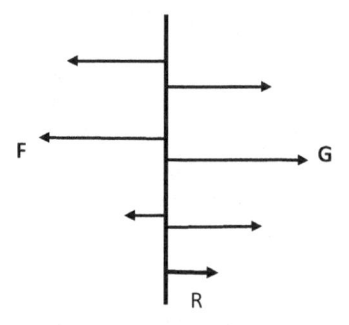

particular patient: in general, we know that symptoms might be caused by the type of diet, emotional stress, etc. However, this general knowledge is not what the model represents.

Note that in Fig. 2.2, it is assumed that the different dispositions compose in a simple additive way, but this is not always or even usually the case. Some dispositions interact in nonlinear ways, and produce synergistic or antagonistic effects. This means that the total effect is greater than (synergistic) or smaller than (antagonistic) the sum of the individual factors. For example, knee pain can be improved or worsened by exercise, depending on the individual context, but also on the amount and type of exercise. We should therefore not expect exercise to be modelled with the same intensity or even direction in two different individual situations, or even for the same individual at two different moments in time. As physiotherapist Matthew Low (2018: 26) notes: 'When evaluating the evidence, one must ask oneself, how does this study relate to my particular patient at this particular time?'. What is relevant for one patient might not be relevant for another.

The vector model, therefore, is a way to *describe the quality* of a causal situation, and not to *measure and quantify* it. Since we are used to thinking about vectors in connection with units of measure, it might take some time to get used to dropping such concerns in this case. But once this is left behind, it should become clear that the vector model allows us to illustrate some central features of causality: different types of causal interference, different degrees of tendency, threshold effects and tipping points, in addition to causal complexity and causal sensitivity. We will present these briefly, one by one.

2.3.6 Two Types of Causal Interference

The effect can be interfered with by removing a vector disposing toward the effect (subtractive interference) or by adding a vector disposing away from the effect (additive interference).

Suppose that in a patient, gastrointestinal symptoms can be counteracted subtractively (Fig. 2.3) by reducing the intake of alcohol, sugar or processed food. But one can also use additive interference (Fig. 2.4), such as probiotic supplements to

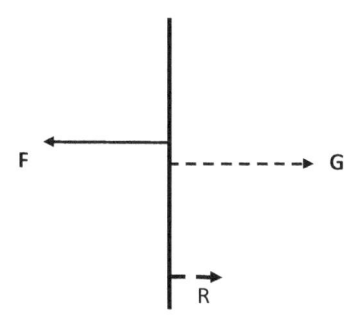

Fig. 2.3 Subtractive interference

Fig. 2.4 Additive interference

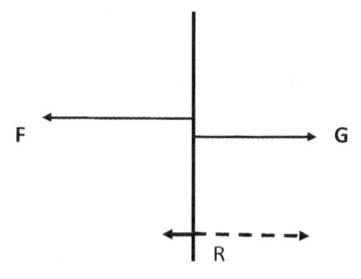

Fig. 2.5 Dispositions with different strength of tendency

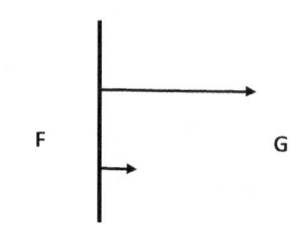

enhance the gut microbiota complexity. Additive interference can be used when subtractive interference is not possible or sufficient to reduce the unwanted effect. Typically, all causal processes can be counteracted by adding something to the situation that tends away from the effect, at least in principle. In fact, most medical treatments are cases of additive interference, and even if an intervention has not been found for all health conditions, the default expectation is that we should keep looking for one.

2.3.7 Degree of Tendency

A disposition has a tendency towards its manifestation with a certain degree or intensity.

All dispositions come in various strengths (Fig. 2.5). For instance, oral contraception has a very strong disposition to prevention ovulation, but also a very weak disposition to produce thrombosis. So even though the correlation between oral contraception and thrombosis looks very weak, statistically, it still counts as causality because there is an intrinsic disposition in the pill toward thrombosis in combination with the appropriate manifestation partners. According to the dispositionalist theory, it is therefore no requirement that a cause produces its effect regularly or even 'often enough' in other similar circumstances, in order to count as causality. What counts is that there is something in the intervention that contributes to the outcome to a stronger or weaker degree.

2.3.8 Threshold Effects and Tipping Points

A threshold effect or tipping point is a stage in the causal process where something conspicuous happens that we might be particularly interested in bringing about or preventing.

The threshold effect (Fig. 2.6) is often a pragmatic and interest-relative matter, but it could also be the point at which a disposition manifests itself into something observable. In medicine and healthcare, a threshold effect might be the stage in the process of an illness where a problem or symptom occurs, such as fever, pain or anxiety attack. It could also represent a crucial stage toward recovery, such as in rehabilitation, where the goals or threshold might change along the way according to changes in the patient and their context.

Thresholds are useful because they can help show whether a situation is close to or far from a tipping point. One patient can be more vulnerable than another, if they are closer to the threshold for illness. In such a situation, a small change in the cause might result in a vast change in the effect. In cases of burnout or chronic fatigue, for instance, the trigger could have been something that might seem relatively harmless from a medical point of view. This could be a conflict at work, an infection or a life-changing event such as a divorce. What triggers an illness is thus not always the main cause of illness, but might simply be the 'straw that broke the camel's back'. In the vector model, one could then illustrate how a small change could have a big impact when the background conditions were already close to the threshold effect, although the same change would not make a difference for a person in a more robust stage of health (see also Price, Chap. 7, this book).

Fig. 2.6 A threshold effect T

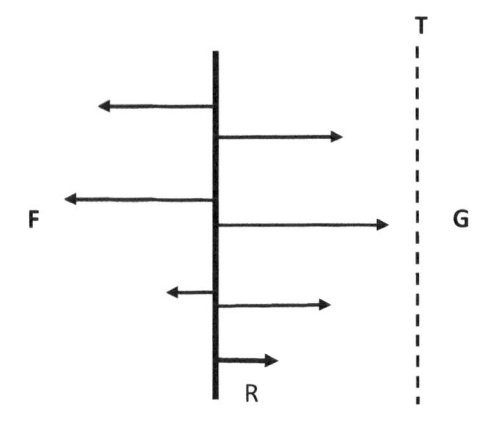

2.4 Philosophy of Causality Influences Scientific Methods

We have seen some main features of causal dispositionalism. This is primarily an ontological framework where causes are seen as dispositions. But how we think about a phenomenon will necessarily influence how we approach it. In this sense at least, ontology has an impact on practice. We will now see how ontological assumptions about the nature of causality influence even the scientific methods used for generating causal evidence.

In the CauseHealth project, we have argued that scientific methods are not philosophically neutral, but carry with them a number of assumptions about the nature of causality (Anjum and Mumford 2018b). How does this work? Let us look at some common methods and see how they attempt to establish causality, epistemologically. From this, we can see what type of features that causality needs to have, ontologically, in order for the method to be a reliable way to test for causality.

Epidemiological and other statistical methods use correlation data to search for causality. They also emphasise large amounts of data and proportion of outcomes. The idea is that more data will lead to more accurate causal conclusions. Ideally, one might think, if we had a complete set of correlation data — past, present, and future — one would also have complete causal knowledge. Philosophically speaking, this fits well with Hume's regularity theory of causality and empiricist agenda. Further, causality is established by observing as many repetitions as possible, where the same cause is followed by the same effect. Some epidemiologists are sceptical of making causal claims based on their observations, and prefer instead to speak of correlations, raised probabilities or relative risk. This meets all the empiricist criteria for not saying something that goes beyond the available observation data and thereby avoiding inductive inference.

Other methods use comparisons of data to establish causality. Comparative methods allow us to search for causes by looking at the difference between two set-ups: one in which the cause is present (test) and a second in which it is absent (control). In randomised controlled trials (RCTs), if the outcome is more frequent in the test group than in the control group, one concludes that the increase is due to the intervention rather than the background conditions, which should be evenly distributed between the groups. The cause is then understood as a difference-maker, as suggested by Hume and Lewis (1973). Crucial for difference-making theory, is that the cause is something that can make a difference to the effect. If no difference can be observed, epistemologically, we have no reason to assume causality, ontologically.

Most scientific methodologies will rely on both regularity and difference-making. In a lab experiment, one compares what happens in the case of intervention with what happens without it, and usually with some repetition. Instead of randomisation of background conditions, these are carefully controlled for. By isolating the cause from interfering factors, one expects to better observe its causal role. Experimental methods also involve an assumption of manipulability, which is the assumption that a causal process can be manipulated in some way. This is crucial when we want to bring about or prevent a certain effect. By manipulating the cause, one also manipulates the effect. This is the basic idea of the interventionist theory of causality, with Woodward (2003) as its main proponent. Note that interventionist theory can be Humean or dispositionalist. If Humean, it would look for whether an intervention makes a difference. If dispositionalist, one would be more interested in the intrinsic properties of the intervention, its causal mechanisms and its influence across individual variations. This tension between statistical and individual effects is also seen in methodological approaches within psychology, as here discussed by psychologist Tobias Gustum Lindstad:

> A prevailing idea among psychologists is that, in order to make psychotherapy evidence based, one has to prove the relevance of therapeutic models and the effects of specific techniques on a group-level. Thus, hundreds of perspectives, models and theories have been thrown into rivalry competing for the best mean results. However, this idea, that the only proper way to uncover the relevant causes is to observe their regular effects (the regularity view) threatens to throw the baby out with the bathwater. Since statistics does not take individual experiences into account, information about aspects that are relevant in each case are lost. Thus, one size does not necessarily fit all, and we must qualify our services locally and individually in any case.
>
> Tobias Gustum Lindstad, 'If statistics don't get me, then what?', CauseHealth blog (https://causehealthblog.wordpress.com/2016/02/05/if-statistics-dont-get-me-then-what/)

2.5 Practical Implications for the Clinic

In this chapter, we have given a brief overview of the dispositionalist theory of causality. We have explained how ontology – how we think the world is – influences epistemology – how we go about investigating the world. We saw that there are ontological assumptions about causality in all scientific methods. But what about clinical practice? What exactly are the practical and clinical implications of understanding causality in the dispositionalist way? We have already mentioned some ways in which dispositionalism might be used as a normative basis for clinical practice.

2.5.1 Causal Evidence Comes from the Patient

Emphasising causal singularism, mutual manifestation and interference, dispositionalism suggests that a major part of causal knowledge will rely on insights into the local context of a unique causal setup. In practice, this means that causal inquiry – both for understanding the condition and deciding how best to treat it – should begin from understanding the full complexity of the patient's situation. This is because the patient will represent most of the causally relevant information needed to understand, diagnose or treat them. The patient's context (the situation, the history, the narratives) is an indispensable source of medical evidence. This is not to say that general theory and population studies are of little use for clinical practice. Rather, local evidence about the patient context is needed in order to make sense of all the other types of evidence and theory available to the clinician. A practice that overlooks it is likely to end up being bad practice, even when it relies on good science and advanced technology. The question arising from this is how a clinician can get a better insight into aspects of the patient's unique history and context that might influence their condition in positive or negative ways. One straightforward answer, which we heard from many of the clinicians we met during the project, is by listening to a patient's narratives, and to the stories they have to tell. This is for instance what physiotherapist Neil Maltby has to say about the matter:

> You're history! Literally. What would you be without it? A void. It is impossible to change it. Your previous choice of job, degree, partner, hobby. Your exposure to family life, upbringing, culture, sports, arts, influential others. Even your genetic make up. Our histories intertwined with previous generation. It would be hard to look at these historical events without acknowledging their causal power in who you are now. What if I had been born into a richer/poorer family, the opposite gender, part of a majority/minority group? Would these not be causally relevant in who we become?
>
> Say we randomly take 100 people off the street and show them the film Terminator. The situation is the same for each person. Same cinema, same time, same popcorn. Will they have the same reaction to the film? Of course not. Because history is more than just events. History is about people. People have dispositions. The best history for me is where people defy their circumstance. This is where we meet personal (or dispositional) attribution.
>
> Dispositional attribution helps explain individual differences to the same stimulus or situation. This is not to say our situation has no impact on us. Clearly it does. It may help shape future dispositions. In life we lean on our internalised dispositions, feelings, previous experience. Two people (or even a single person with a time gap) may internalise the same situation in very different ways. This means we cannot rely on humans reacting robotically especially to complexity.
>
> Is this even important?! Well I'd say so. It means as healthcare clinicians we are not striving for uniformity in treatment (situational attribution) because, as research shows, not everyone will respond to this. I think there is a lingering hope that one day we will come up with perfect protocols for lower back pain, fibromyalgia, tendon pain, irritable bowel syndrome and depression. This seems to be the aim in most research I read. Treatments based on pathology tend ultimately to look at situational attribution and not the dispositions of the individual.
>
> Neil Maltby, 'You're history (hasta la vista, baby)', CauseHealth blog (https://causehealth-blog.wordpress.com/2016/02/19)

2.5.2 There Is No Standard or Average Patient

Assuming causal singularism and causal complexity, dispositionalism suggests that there might never be two identical causal situations in practice. All patients are in some sense medically unique, with different genetics, life situation and biography.

In practice, this means that we should not expect that there is a 'normal', 'ideal' or 'standard' patient or even a normal response to a treatment. If something happens in one patient that cannot be backed up statistically or observed in other patients, this does not rule out the possibility of causality. Causal singularism means that all causal processes are intrinsic and particular. Effects happen in the single patient, as the result of multiple dispositions, many of which are unique to that patient. Physiotherapist and researcher Wenche Schrøder Bjorbækmo writes about standardisation:

> At the end of the 1990s I perceived standardised testing and standardised procedures as the future. As a tester I became concerned with performing the tests correctly, which meant following the standardised procedures. The tasks I asked the children to do, and the questions I asked the parents were guided by the instructions and the structure of the test. It was important to remember the order of the tasks and questions. Several of the tests had many tasks and questions, and there wasn't just one test to learn, but many. Each test had been adapted for different purposes and for different patient groups.
>
> The test directs the professional view in particular directions, and thus away from anything else. When something in this way is brought into the foreground, other aspects, of for instance a child's functioning, are disappearing into the background. In many ways this experience led me to think that in testing I was actually gaining less knowledge about the child and his or her functioning than in traditional clinical observations.
>
> This experience of having 'seen' less and received less information frustrated and disturbed me. The test "demanded" a special form of communication and being together. I experienced the standardised framework of the test and administrative demands as framing the communication and relation made possible between the child, parents and myself.
>
> Wenche Bjorbækmo, 'Glasses and blind spots: through the eyes of a tester, CauseHealth blog (https://causehealthblog.wordpress.com/2017/11/20)

2.5.3 Unexpected Outcomes Are Valuable Causal Lessons

Dispositionalism stresses that all causal processes can be counteracted, subtractively or additively. So even when the effect typically follows from the cause, it is still possible to have the cause together with some interferer that is preventing the effect from happening. When this happens, one should aim to understand the dispositions involved and the causal mechanisms by which they interacted to produce or inhibit the effect.

This type of causal knowledge is particularly important for predicting how a patient will benefit from a treatment, or if there are any risks involved. While all treatments have a targeted effect that is tested and established, one can learn something important about the treatment's other dispositions from their side effects. Side effects are often rare and unexpected, yet they point to dispositions in the patient that were able to causally interact with the drug. From this, we can develop new causal hypotheses for theory development, which are also relevant for basic research within medicine and biology (Rocca et al. 2019; Rocca et al. 2020). Ivor Ralph Edwards, medical doctor and senior medical advisor of the WHO collaborating Uppsala Monitoring Centre for International Drug Monitoring, has been dealing with the problem of detecting unexpected effects of pharmaceutical interventions for decades. In the following text, he explains why thinking of causes as dispositions can be useful in pharmacovigilance:

> There is an ongoing debate about how to analyse and evaluate the data we gain from large data sets, and particularly what we can say about causality – after all, one is bound to find correlations by chance in vast amounts of data and with multiple analyses; but all this assumes that causality is a linear process which can be evaluated epidemiologically. Causality in real life, however, is usually multifactorial and complicated, and pharmacovigilance is concerned with data from complex healthcare systems in which multiple interrelating factors evolve. The data we collect is affected by those changes over which we have no control....
>
> One different approach to causality in pharmacovigilance is causal dispositionalism and is applicable to complex data. This approach considers the innate characteristics (the dispositions) of both the medicinal product and the exposed patient – some property, state, or condition that, under certain circumstances, gives the possibility of some further specific state or behaviour. The relevant properties of the medicines would include its various pharmacological actions (pharmacodynamics), its distribution in the body (pharmacokinetics), and its interactions with other drugs. The relevant properties of the patient would include specific susceptibilities, such as genetics, age, sex, physiological state such as body weight or pregnancy, co-morbidities, drug-drug interactions, and social and environmental factors that have affected the patient.
>
> Consider a medication M, with a set of dispositions, $M[d_1]$, $M[d_2]$, and so on, known to be able to cause benefits and harm, and patient P with dispositions $P[d_1]$, $P[d_2]$, and so on. We may then begin to investigate the probabilities that any $M[d]$ will produce beneficial or adverse outcomes in a patient with any $P[d]$, asking the questions 'how?', and 'when?', using whatever information we have about the medicine M and the patient P to determine the benefit to harm balance. This type of analysis is not merely probabilistic, but also takes into account the strength – the power – of M to affect P, as well as any outside factor that interacts with the causal link, e.g. drug-drug interactions. It also explicitly takes into account the power of P to respond to M. A disposition may be present but not become manifest until its power reaches a particular threshold, e.g. above a certain dose of a medication, in combination with, for example, a certain degree of renal function impairment in the patient.
>
> Alternatively, a medicine with disposition $M[d_4]$ may have maximal effects in patient P_1 with dispositions $P_1[d_{2,3,4,5,6,7}]$, partial effects in Patient P_2 with dispositions $P_2[d_{2\&7}]$, and partial or maximum effects in other patients P_n with, say, dispositions $P_n[d_{2,3,4,5,6,7,8,9}]$, but only in certain environment where the two extra dispositions d_8 and d_9 result in an additional influence, such as might occur when syncope from a vasodilator only happens when a susceptible patient is dehydrated...
>
> Let's be very broad-minded about what new value we can find in the multiplicity of big real-life data sets we can utilise to examine benefit and risk and thereby improve therapy. (Edwards 2018: 28–29)

2.6 To Sum Up...

In this chapter we have argued that a genuine revision of the norms and practices in clinical work needs to start from a revision of the way we think about the world, and in particular the way we think about the most foundational concepts, such as causality. We have presented the dispositionalist theory of causality and explained why this theory is better suited for the clinic than the orthodox Humean theory which motivated the EBM framework. We have argued that causality ought to be understood as something singular and intrinsic rather than as a pattern of regularity across different contexts. From a dispositionalist perspective, causal knowledge ought to start from the single case. In the clinic, this means that the more we know about the multiple dispositions that are involved and how they interact in this specific context, the better equipped we will be to make good and relevant explanations, predictions and decisions for the individual seeking care.

References and Further Readings

Anjum RL, Mumford S (2017) Emergence and demergence. In: Paoletti M, Orilia F (eds) Philosophical and scientific perspectives on downward causation. Routledge, London, pp 92–109

Anjum RL, Mumford S (2018a) What tends to be. The philosophy of dispositional modality. Routledge, London

Anjum RL, Mumford S (2018b) Causation in science and the methods of scientific discovery. Oxford University Press, Oxford

Cartwright N (1989) Nature's capacities and their measurements. Oxford University Press, Oxford

Edwards R (2018) Living with complexity and big data. Uppsala Rep 78:28–29

Gillies D (2018) Causality, probability, and medicine. Routledge, London

Howick J (2011) Exposing the vanities – and a qualified defense – of mechanistic reasoning in health care decision making. Philos Sci 78:926–940

Hume D (1739) A treatise of human nature. In: Selby-Bigge LA (ed) Clarendon Press, Oxford, 1888

Lewis D (1973) Causation. In: Lewis D (ed) Philosophical papers ii. Oxford University Press, Oxford 1986:159–213

Low M (2017) A novel clinical framework: the use of dispositions in clinical practice. A person centred approach. J Eval Clin Pract 23:1062–1070

Low M (2018) Managing complexity in musculoskeletal conditions: reflections from a physiotherapist. In Touch 164:22–28

Martin CB (2008) The mind in nature. Oxford University Press, Oxford

Mumford S (1998) Dispositions. Oxford University Press, Oxford

Mumford S (2004) Laws in nature. Routledge, London

Mumford S, Anjum RL (2010) A powerful theory of causation. In: Marmodoro A (ed) The metaphysics of powers: their grounding and their manifestations. Routledge, London, pp 143–159

Mumford S, Anjum RL (2011) Getting causes from powers. Oxford University Press, Oxford

Rocca E (2018) The judgements that evidence based medicine adopts. J Eval Clin Pract 24:1184–1190

Rocca E, Copeland S, Edwards IR (2019) Pharmacovigilance as scientific discovery: an argument for trans-disciplinarity. Drug Saf 42:1115–1124

Rocca E, Anjum RL, Mumford S (2020) Causal insights from failure. Post-marketing risk assessment of drugs as a way to uncover causal mechanisms. In: La Caze A, Osimani B (eds) Uncertainty in pharmacology: epistemology, methods and decisions, Boston series for the history and philosophy of science, Springer, Dordrecht

Russo F, Williamson J (2007) Interpreting causality in the health sciences. Int Stud Philos Sci 21:157–170

Woodward J (2003) Making things happen: a theory of causal explanation. Oxford University Press, Oxford

Open Access This chapter is licensed under the terms of the Creative Commons Attribution 4.0 International License (http://creativecommons.org/licenses/by/4.0/), which permits use, sharing, adaptation, distribution and reproduction in any medium or format, as long as you give appropriate credit to the original author(s) and the source, provide a link to the Creative Commons license and indicate if changes were made.

The images or other third party material in this chapter are included in the chapter's Creative Commons license, unless indicated otherwise in a credit line to the material. If material is not included in the chapter's Creative Commons license and your intended use is not permitted by statutory regulation or exceeds the permitted use, you will need to obtain permission directly from the copyright holder.

Chapter 3
Probability for the Clinical Encounter

Elena Rocca

3.1 Uncertainty and Probability in the Single Case

'Healthcare professional' are words we use to describe a group of professionals with a variety of different specialisations, approaches and backgrounds. And yet, one can draw out some commonalities from this diversity. There are some aspects that everyone working through consultation with suffering individuals will recognise, in different degrees, as a daily part of their job. One of these aspects is uncertainty.

No matter their specific field of expertise, every healthcare professional must cope with the fact that all patients are different. Not only is every biological setup unique; the multiplicity of contextual conditions, lives, habits and stories make every patient a special case. Because of this, the practice of inferring information from one patient (or group of patients or experimental model) to another patient always entails some unknown margin of error. Uncertainty is intrinsic to all the phases of the encounter with the single patient, from diagnosis to prognosis to treatment. Some of these uncertainties must be accepted as such, but most of them need to be somehow qualified. Even the most experienced healthcare professional needs, in every single instance, to find an answer to the question:

What is the probability that this intervention will work *this* time, for *this* particular patient?

This question, we can say, is universal in the clinical encounter. However, there are different ways to approach it, conceptually and philosophically.

We are used to think about probability as *the* way to deal with uncertainty. Probability turns uncertainty into something more tangible by somehow giving it a qualification, or even a quantification. For instance, we say that the chances that a

---

E. Rocca (✉)
Centre for Applied Philosophy of Science, School of Economics and Business, Norwegian University of Life Sciences, Aas, Norway
e-mail: elena.rocca@nmbu.no

© The Author(s) 2020
R. L. Anjum et al. (eds.), *Rethinking Causality, Complexity and Evidence for the Unique Patient*, https://doi.org/10.1007/978-3-030-41239-5_3

certain treatment will work in the single patient is 30%, fairly low or higher compared to another treatment. This will inform the clinical choice in a more satisfactory way than just acknowledging that "the outcome is uncertain". Thinking in terms of probability is then a useful tool for dealing with uncertainties for the single patient. But what do these numbers mean? Can they be interpreted in only one way, or in many?

Thinking in terms of probability offers multiple ways, rather than only one, for dealing with uncertainty. Although we might assign similar probabilities to a certain event, there are different ways to interpret what such assigned values, or descriptions, mean. Think, for instance, about a healthcare professional who assures a patient that the probability of recovery with a certain intervention is 'very high'. How should the patient understand such a statement? That the majority of the patients, previously treated with the same intervention, finally healed? That the professional has a high confidence in the positive outcome of this particular case? Or does it mean that the patient's condition at this moment is optimal to respond to this particular treatment, considering its mode of action? Clearly, these are three different types of statement, and might lead to different clinical decisions.

In this chapter, we will first explore the basic assumptions that hide behind the usual conceptualisation of probability in evidence based approaches. One version of probability is objective and the other is subjective. We then explore a dispositionalist concept of probability, which we think is most relevant for the clinical context. We will show how this understanding of probability can help integrate evidence from healthcare guidelines with evidence from the single patient.

3.2 Probability from Statistics: Frequentism

To understand the frequentist approach to probability, imagine that you see a coin for the first time, and you notice that it is two-sided. You wonder what is the probability of the coin landing heads when tossed. One way to approach this question is to toss the coin many times, and count how frequently you get a head. After a sufficient number of repetitions you might calculate that about half the outcomes were heads, and therefore you can infer that the probability of getting a head at the next toss is ½ or 50%. Crucially, you would not be confident in drawing a conclusion after only 3, 5, or 20 tosses. The more instances you have to base your calculation on, the more you can trust the result to be accurate.

We see that the frequentist approach calculates the probability that a certain event will happen by investigating how often it happened in the past. Philosophers call this type of approach 'empirical', meaning that it is exclusively based on observation. Recall that David Hume only trusted knowledge that could be observed through our senses, which was his empiricist starting point (see Anjum, Chap. 2, this book). To know how probable it is for the coin to land heads or tails, you would then not need to understand anything about the coin's properties or hidden dispositions. You simply toss the coin as many times as you can and count the distribution of outcomes.

In reality, the calculation of past events is usually much more elaborate than in the coin example, which is why scientists use statistical models and tools to calculate the relative frequency of a certain outcome in a sequence of events.

Note also that this approach was developed by thinking about games of chance, similar to coin tossing, and as such it needs two important premises to count as successful. First, the frequentist approach presupposes that there is the possibility (at least theoretically) of an infinite number of repetitions. Second, one needs to repeat many instances of the exact same conditions. As a consequence of this, it is not possible to calculate a frequentist probability for a single case if it cannot be repeated. From a single case, all we know is the actual outcome, not the proportion of such outcomes over a series of similar trials.

> Simply put...
>
> Frequentism means to calculate the objective probability that a certain event will happen based on the proportion of positive outcomes in a sequence of trials. To calculate probability, one then needs to observe how often the same type of event happened in previous, similar cases.

3.2.1 Frequentism and Evidence Based Approaches

Now imagine that, instead of the coin toss, we are talking about an intervention for the single patient. Clearly in this case, we are missing both premises that we saw are important for the frequentist approach to be reliable. We only have one instance of this particular patient meeting this particular intervention under some particular conditions. How can a frequentist notion of probability possibly be applied to the single patient? How can we say something about the probability of effect for this individual case?

There is a way out, and it is one that is widely used in evidence based approaches. By assuming that each single patient is a statistical average of a group of individuals that are similar enough to the patient in question, one can use the group as a representative for that patient. The probability that an intervention works for this patient can then be derived from the calculation of the statistical frequency of successful outcome in their patient group. This is the principle on which clinical studies are based. To say that a patient has 30% probability of recovery from a certain intervention based on clinical studies, means that 30% of sufficiently similar patients who tried that intervention under sufficiently similar conditions, recovered (at least for the patients who participated in those trials). These kinds of predictions are common in evidence based medicine and practice, which takes the best evidence to be statistical evidence from clinical trials.

From a healthcare perspective, however, there are some problems with this type of reasoning. By seeing the patient as an average of similar patients, one must be able to define what counts as 'similar'. Which pieces of information are relevant in each case? Similar age, medical history, lifestyle or social status? We are normally not aware of which factors play a causal role in the single process, which is why it can be misleading to see a patient as an average of groups of other patients, even within the appropriate patient group. This issue is well acknowledged, and it is sometimes referred to as the 'reference class' problem. The reference class problem influences the interpretation of statistical data from population studies for the purposes of inferring the probability of an outcome for the single patient. Imagine for instance having to calculate the probability of a patient to respond to a certain class of anti-depressants. The patient has countless properties (woman, young, history of eating disorders, wealthy, diabetic, highly educated, hyperactive…), but which of these will have a role in her condition and in the therapeutic process? This is known only partially. Our patient is a member of many different classes (young wealthy women, diabetic patients, patients with eating disorders), for which the frequency of recovery from the anti-depressant differs. However, it is not obvious how the patient should be 'classified' in relation to her depression and its treatment, since we do not have a complete knowledge of which of her properties will play a causal role in her clinical development. A number of tactics have been suggested as a solution to this problem, and many of these consist in moving away from a purely frequentist approach to probability, and including different types of evidence in the calculation, such as mechanistic evidence (Clarke et al. 2013, 2014; Wallmann and Williamson 2017).

3.2.2 Randomisation, Inclusion Criteria and Exclusion Criteria in Population Trials

Let us now look at the case of clinical trials, of which randomised controlled trials (RCTs) are currently considered the most reliable. RCTs have the purpose to assess the frequency of recovery in a group of patients that received a treatment, compared to a group of patients that received only a control or placebo. Based on such frequency, one can predict the probability for the treatment to have a positive outcome for the single patient. This is a frequentist approach to probability, as we described above. Recall the example of the coin toss. There are two important premises for being able to infer the probabilities for the next toss, from the frequency of outcomes of previous coin tosses: first, the repetitions should be many, and second, they should all happen under similar conditions. How can these conditions be met in an RCT?

The first requirement is served by including a large number of patients in the trial. One criterion of quality for an RCT's design is that the more patients included, the better. This is however not sufficient. If we want to end up with many instances in which the same treatment is tried out in similar contexts, we also need to cancel out the influence of individual variations between different patients. This is pursued with two strategies at the same time.

The first strategy is *randomisation*. The reasoning goes like this: if one randomly assigns the patients to the two groups (the group of patients who receive the treatment and the group of patients who receive the placebo) and the groups are large enough, then there is greater probability that the relevant causal factors are distributed evenly across the groups. This way, the two groups can be considered homogeneous, or at least similar enough. This is important in case we detect a statistical difference in the outcome among the two groups, since it allows us to infer that any such difference is caused by the intervention we are testing, rather than by some other factor. Note that the difference spotted in an RCT is at the population level, not in single patients, since there will be many individual variations within both the test group and the control group.

There is an additional strategy for cancelling out the influence of individual variation on an RCT. This is to define in advance strict *inclusion and exclusion criteria* for selecting which patients qualify to take part in the study. Typically, the patients included in the study belong to a certain age range and have a certain medical profile.

When designing an RCT, it is important to define in advance inclusion and exclusion criteria, so that the sample included in the study is representative enough for the population we want to study. In other words, at the end of the study one wants to use the results observed in the selected sample in order to say something *general* about the population we wanted to study. If the aim of the RCT is to get some information about Norwegian women of fertile age, for instance, the study sample should not be imbalanced with respect to age, geography, income or health condition. As a result of this knowledge, exceptional cases, outliers, patients with conditions that could influence or confound the interpretation of the statistical results, or patients at higher risk of adverse effects might be excluded.

3.2.3 *Internal and External Validity of Causal Claims from Randomised Controlled Trials*

These two strategies (randomisation and predetermined inclusion and exclusion criteria) are aimed at increasing the reliability of the trial's results. We want the study to allow us to detect the effect of the intervention, and not to be confounded with other factors that could influence it. This is also called the *internal validity*, or *reliability*, of a causal claim based on a certain study design. A different matter, however, is to figure out what use we may have for such causal claim, once we know they are valid for the experimental sample *on average*. How does the knowledge of how often something happened to the participants of an RCT apply when predicting what is going to happen in the single patient? This is the question of *external validity*, or *relevance*, of a causal claim based on the study for the case in question. The external validity of causal claims based on RCTs might be low when we are faced with marginal cases, or with multi-morbid, chronically sick patients, who rarely meet the inclusion criteria of an RCT. To what extent does the available evidence from clinical studies represent these patients? Scientists and philosophers of science have worried about this issue, when thinking about evidence based decisions and how they should be made. As we shall see, the inductive inference from 'it worked there' (in the study) to 'it will work here' (in my case) is not an easy one, and it is paved with challenges and pitfalls where the doctor's expertise and knowledge of her patient seem to be indispensable ingredients (for a critical discussion of the external validity of RCTs, see Rothwell 2005, 2006; Cartwright and Hardie 2012).

3.3 Probability as Degree of Belief: Subjective Credence

Say you are evaluating whether to prescribe a painkiller to a 30 year old patient and want to predict the probability that the patient will get gastrointestinal side effects from the treatment. Imagine that you have also already prescribed painkillers of the same class to 100 patients, and of those patients, 30 patients experienced that side effect. If this is the only information you have on the matter, and you want to make a prediction about the probability of side effects for your patient, you might think that such probability is 0.3 or 30%.

Let us assume that after a conversation with your patient, you learn that she suffered from chronic gastritis for a number of years and only got better 5 years ago, while she still suffers from temporary relapses. With this information at hand, you might now change your belief on the likelihood of a gastrointestinal effect in the patient, from 0.3 to >0.3. Another colleague in the same situation, however, might just have been to an information meeting with the manufacturer of the pain killer and learned that this particular painkiller acts through a different molecular pathway than the others in the same class, and does not interfere with gastrointestinal pathways. In light of this additional information, your colleague might have a different opinion than yours, and conclude that the probability of the patient getting a side effect is not that high after all, and at <0.3 or even less.

We see that, within this philosophical theory, we can have three different estimates of probability for the same patient in the same situation. The estimate will depend on which relevant facts we are aware of and how important we think these facts are for this particular patient. This suggests that the estimation of probability P is not objective or ontological, but subjective and epistemological: it concerns the information and knowledge that the healthcare professional has available at that time. The value assigned to P will then be the subjective measure of one's own degree of belief that an outcome O will happen, given the available evidence E. In mathematical language, this can be written in the following way:

P (O|E) ('the probability P of the outcome O happening, given the evidence E')

Every time we acquire new information, the evidence E changes and therefore our degree of belief in the outcome is updated. This might strike us as an intuitive and straightforward practice, but there are nevertheless some practical and philosophical issues to consider.

3.3.1 Updating Belief

The first issue is a practical one. How exactly should we update our degree of belief in light of new information? Who is apt to do it? And does 'subjective measure' of probability entail an uncontrolled subjectivism, by which anyone and anything goes? Certainly not. Proponents of subjective probability postulate that the way in which new evidence is used to update the degree of belief must follow some common rules. These are the rules of probability calculus. In other words, two clinicians might calculate a different probability of a certain treatment to work in a specific patient, but this will only happen because they have access to different information. But the *way* in which a piece of evidence updates the belief, should be the same for both clinicians.

> Simply put...
>
> Subjective probability (credence) is the degree of belief, or confidence, in a specific outcome given certain available evidence, as estimated by a suitable agent. A suitable agent is an agent that uses the rules of probability calculus in order to update its expectations, and the value of the probability is always expressed quantitatively.

One way to update beliefs or expectations in light of new evidence is given by the Bayesian formula. The Bayesian formula for calculating probability includes some *prior* probability (or belief), which in light of a new piece of evidence is then updated into *posterior* probability (or belief). One necessary assumption of the Bayesian formula is that we adopt a specific way to calculate probabilities that depends on the value of another probability. In the clinic, the need for such evaluations is quite common. For instance, we might want to know the probability that a therapy works well, given the patient's conditions. This is called 'conditional probability', which intuitively means 'the probability of an outcome given an intervention'. In probabilistic calculus, however, 'conditional probability' has a technical meaning and is calculated in a specific way. (For more details on the notion of conditional probability, see Anjum et al. 2018.)

3.3.2 Understanding the Basic Bayesian Formula

Bayesian calculations of probability can be rather complicated, and often they are made through a computational tool, such as a software. Many of the software available to the decision makers, not only in medicine and healthcare, but also more generally in the field of risk assessment, are based on Bayesianism. These software programs calculate the posterior probability of an outcome, every time new evidence is typed into the software. The disadvantage of making decisions based on software packages is that the user has to adopt the assumptions of the programmer, without the possibility of critical consideration. Although the programming of software based on Bayesian principles can be complicated, the principle on which the whole system is based is not difficult to grasp. Let us have a look at it.

The Bayesian formula postulates a way to derive posterior probabilities from the combination of prior probabilities, new evidence, and the likelihood of the event that constitutes the new evidence occurring if our prior hypothesis is correct. In its most basic form, the formula looks like this:

P(Hypothesis | Evidence) = P(Hypothesis) x [P(Evidence | Hypothesis) / P(Evidence)]

First of all, let us explain every term of the formula with an example from the clinic.

P (Hypothesis) is the prior probability, or the probability of a hypothesis being true prior to getting to see the new evidence. For instance, P (Hypothesis) could be the probability of a patient having hypertension. Let us say that the patient is young and has a healthy lifestyle. The prior probability in this case is low. Note that the problem of how to assign prior probabilities is an important one, and Bayesians disagree on the matter. We will come back to this later on.

P (Hypothesis | Evidence) is the posterior probability, or the probability that the Hypothesis is true given that we get to know the new Evidence. In our example, this could correspond to the probability that the same patient has hypertension after she came to consultation complaining about a severe headache (Evidence = headache).

P (Evidence | Hypothesis) is the likelihood that the new Evidence occurs given that our Hypothesis is true. In our example, it would be the likelihood that our patient has a headache given that she has hypertension.

P (Evidence) is the likelihood of the new Evidence happening at all. For instance, the likelihood of a young healthy person having a severe headache.

Now that we know what the terms mean, let us have a look at what the formula tells us.

In order to obtain the posterior probability, Bayes is telling us to multiply the prior probability by the factor [*(likelihood of Evidence given Hypothesis)/(likelihood of Evidence at all)*]. Why this?

The first observation is that the posterior probability is directly proportional to the likelihood of evidence happening given that the hypothesis is true. We can understand this by thinking about our example. The likelihood of a hypertensive patient having a severe headache is high, therefore the posterior probability (i.e., our degree of confidence) of the patient having hypertension after knowing that he has headache, is higher than before knowing it. Let us imagine for a moment that the patient, instead of complaining about a headache, had complained about lower back pain. The probability of a hypertensive patient having lower back pain is not particularly high, therefore the value of posterior probability would not be higher than the probability prior to knowing that he has back pain.

The second observation is that the posterior probability is inversely proportional to the probability that the new evidence happens at all. In other words, the lower the probability of the new evidence, the higher is the updated belief on the hypothesis. Why? Think again about our example. P (Evidence) corresponds to the probability that a young healthy patient has severe headaches. This probability is low. Therefore, if it happens in our patient, it updates consistently the belief that we had in the hypothesis that he is hypertensive, before knowing the new piece of information. But let us now imagine that the patient, instead of complaining about a headache, complains about moderate fatigue. Moderate fatigue is a relatively frequent condition even in young and healthy people. Therefore, after knowing that the patient is often tired, our posterior belief in the hypothesis of hypertension is not that much higher than it was before knowing the new evidence.

We see, then, that the Bayesian formula is intuitive, as long as we bear in mind that prior and posterior probabilities are not intended as existing entities, but rather as *subjective* knowledge, or *degrees of belief*. Note that, ontologically speaking, it would not make so much sense to suggest that the probabilities of me having hypertension given that I have a headache somehow depends on the probability of a generic healthy woman having a headache. (For a dispositionalist discussion of the Bayesian formula, see Anjum and Mumford 2018, ch. 19 and 21.)

There is, however, something called *objective* Bayesianism, which might create some confusion. For our purposes here, it is sufficient to point out that what we said so far is general enough to apply both to subjective and objective Bayesian inference. The difference between these two is the strategy one might use to assign prior probabilities: what is the probability of a young healthy woman having hypertension, if we just know about her that she is young and otherwise healthy? How should one assign such probability? Objective Bayesian inference postulates that there needs to be a rational and agreed way to do this task (e.g. Williamson 2010). For instance, one could use the incidence of hypertension in the general population of young and healthy women.

3.3.3 Uncertainty as Lack of Knowledge

One important aspect of interpreting probability as degree of belief is to notice what it is exactly that generates the uncertainty. Given a certain patient and a certain treatment, why are we uncertain about the outcome? According to this philosophical understanding of probability, any uncertainty in prediction comes from lack of knowledge. There are many sources of uncertainty in the clinic. We might lack information about our patient, her condition or about how the treatment works. We might base our expectation on clinical studies or lab tests that are flawed. Not least, we cannot know the complete set of possible outcomes, for instance whether the treatment might provoke some hitherto unknown side effects in this specific patient. If all the possible knowledge were available to us, there would be no uncertainty left.

Within the subjective Bayesian notion of probability, uncertainty is treated as an epistemological matter: a matter of what we can possibly know. In this case, probability is not understood as an ontological matter. The uncertainty or degree of belief that we estimate using the subjective Bayesian inference should then not be understood as something that exists outside of us, in the world. In other terms, given a certain set of initial conditions (a patient, an illness, a stage of illness, a treatment, a context of treatment) there would be no inherent uncertainty about the outcome. The possible outcome is only one: the trouble is that it is impossible to know it for sure unless we were omniscient beings. This suggests that the *credence* notion of probability assumes that causality is a deterministic matter: of all (probability 1) or nothing (probability 0). Probability itself, then, does not come in degrees. Probabilistic claims do not, therefore, express something about the causal strength of an intervention, but about the limits of our knowledge and our confidence in a particular outcome. We will now look at a third notion of probability that instead sees probabilities as ontological, dispositional and intrinsic.

3.4 Probabilities as Dispositional and Intrinsic: Propensities

We have seen two possible interpretations of probability: probability as the frequency of outcome in a relevant population, and probability as subjective degree of belief, or credence. Both these approaches, however, might seem somehow inadequate or unsatisfactory for the clinic. Frequentism because it must treat the patient as a statistical average of their relevant sub-group, and credence because it takes probabilities to be entirely subjective. Indeed, clinicians are likely to think of the probability that their patient will recover as ontological: something real, physically existing in the world, but also to some extent as something that is intrinsic to the patient. Regardless of whether I have a limited knowledge about a certain condition and its prognosis, there is an actual, existing probability that the patient will recover. And this is entirely independent from my own subjective belief. Such physical probability that the patient will recover is produced by the patient's own pathophysiological and contextual situation and can be called a *propensity*.

There are many different understandings of propensity in literature. Common to all these definitions, however, is that propensities refer to the single event. Karl Popper, one early proponent of the propensity theory of probability, describes propensities as dispositional properties of singular events (Popper 1959). Propensities, we might say, are an explanatory understanding of probabilities. The probability of me falling asleep after an injection of morphine is explained and qualified by the sedative power of morphine, by my degree of habituation and by other properties of the whole situation, which in total we can call an overall propensity. A further example might help to clarify this idea.

An elderly person affected by influenza, for instance, has a certain probability of recovery, which is generated by the type of viral infection, the state of the patient's immune system, his general health and his context and lifestyle. This set of dispositions of the whole single situation generates a certain propensity of the patient to heal. Certainly, a clinician might draw insights into this individual case from the previous experience of similar cases. However, the probability of recovery of this particular patient is affected only by the physical properties, or dispositions, in place (type of virus, patient's immune system, et cetera). Ontologically, this probability is independent of the outcomes in other similar cases. Epistemologically, what happens in similar cases can be an indication for the probability of outcome in the single patient, but not necessarily. To use a brutal example, if a bus full of patients with a rare condition crashes on its way to a medical conference, their collective death does not affect the propensity of the patient who missed the bus to survive the rare condition. On a frequentist account, however, that patient's probability of survival will have changed with the bus accident (without it being a good epistemological indicator either, in this case).

> Simply put…
>
> The individual propensities of a single patient, treatment or context are given by its unique combination of dispositions and the dispositions' degree of tendency toward certain outcomes *in that individual case*.

3.4.1 Individual Propensities Are Not Always Seen Through Frequencies

A dispositionalist understanding of causality, as described in Chap. 2, fits best with a view of probabilities as single *propensities*. Although propensity interpretations of probability are less known than frequentism and credence, it has been defended by philosophers (e.g. Popper 1959, 1990; Mellor 1971; Gillies 2000, 2018) and scientists (e.g. Bohm 1957) since the early 1900s. Propensities seem particularly relevant in the clinic.

Think about an epileptic child who is about to start a therapy with valproic acid, a widely used anticonvulsant. What are the probabilities that the patient develops liver toxicity as an undesired effect of the drug? One possible answer is that the child will most probably be unhurt, thus with probability close to zero, on the basis of the outcome for the majority of children. However, this evaluation might be met as superficial, or unsatisfactory, given that children *can* be hurt by valproic acid, sometimes even fatally.

A better clinical approach to the question might be: what are the *propensities* of *this* child to develop liver toxicity from valproic acid? If the child has the particular physical and contextual onset that makes her sensitive to the drug's undesired effect, then the toxic outcome, no matter how rare, will nevertheless be very likely for her. The frequency of the toxic outcome in other children can in some cases be indicative of the individual propensity to the outcome (for instance, if we consider closely the properties of the patient in comparison with the properties of the harmed children), but is not necessarily so.

3.4.2 Propensities as Qualities

How should propensities be expressed? Can a number between 0 and 1 be estimated for this purpose, as is generally done in probability theory? Philosophers have different views on this matter, depending on their understanding of propensity. In the CauseHealth project, however, we favour a singular and qualitative, rather than a numerical, description of propensities. (For an approach to propensities that is more compatible with frequentism, see Gillies 2018.)

Propensities, we said, are generated by dispositions. These are intrinsic qualities of things. But a propensity also depends on the disposition's magnitude or intensity, which, although being in some sense quantitative, cannot be directly derived from statistical frequencies. The presence (or absence) of mutual manifestation partners for a certain disposition, and the presence (or absence) of possible causal mechanisms which might result in an outcome, affect the propensity of such an outcome to happen. Evaluating the propensity for an outcome therefore requires describing and understanding, at least partially, *how* that certain outcome could or could not happen. Numbers or scores cannot completely fulfil this purpose, at least not alone, since what happens statistically at a population level might not reflect what happens in each individual case.

On average, for instance, a population might seem to have a weak propensity toward cirrhosis, but this average only represents the sum of all the manifestations of individual propensities toward cirrhosis. Individual propensities will depend on a number of dispositions related to age, gender, genetics, lifestyle, diet and medical history. All of these will be different from one individual to another, so we should not expect two people to have exactly the same combination of dispositions, or dispositions with exactly the same magnitude.

3.4.3 Propensities and Prediction

There is a further important consequence of propensities being generated by dispositions. We saw in Chap. 2 that dispositions can exist unmanifested. A patient might carry a certain genetic mutation that disposes toward an allergic reaction from penicillin, for instance. However, we are likely to remain unaware of such a disposition until it meets its proper mutual manifestation partner. In other words, it might be difficult to correctly evaluate the propensity toward this toxic reaction until the patient uses penicillin. In some cases, we might even get it totally wrong. As a consequence, evaluations of propensity ought to be carried out with some degree of epistemic humility. That means, there might be dispositions and interactions in this particular case, which we were not aware of at the moment of evaluation. Any evaluation or prediction must therefore be interpreted with some caution.

Although this attitude is valid for any interpretation of probability, it is especially important when we think about propensities and dispositions. Accordingly, probabilities as propensities are a matter of qualitative evaluation, theoretical knowledge and practical expertise, and cannot be generated by an algorithm as a definite number. Notice that this does not mean that propensities are more fallible than other ways to calculate probability. Rather, it might just make us more aware of the fallibility of prediction.

If we accept the dispositionalist notion of probability, a question remains: how should clinical inquiry (as well as research) be organised in order to uncover propensities for the single patient?

3.5 Propensities and the Clinic

A clinician who adopts the propensity approach to probability also adopts a certain specific approach to clinical inquiry. In this section, we list some of the methodological and epistemological implications of the propensity view of probability. Note that these follow from the particular version of propensities here presented and dispositionalism presented in Chap. 2, and that other versions of propensity theory might have other implications.

3.5.1 The Importance of Local Knowledge

The more one knows about the dispositions and interactions in place in the particular case of interest, the more reliably one can evaluate the propensity toward one specific outcome in that case. This might sound like nothing particularly new. The frequentist approach, indeed, also requires us to know as much as possible about the case of inquiry, so that the most relevant sub-population can be found, and more reliable statistics performed. So what is particular about the propensity approach? The difference is in the *type* of knowledge required. Uncovering propensities requires knowledge about local processes and interactions, rather than knowledge of mere values and parameters. Typically, local processes and interactions need to be observed in their own context and as aspects of a whole, while values and parameters can be picked and chosen, and analysed in isolation. Local knowledge of a patient, ideally, would not be reduced to knowing the value of his biomarkers or genetic onset, but requires that we have as much knowledge as possible about his unique context, including history, lifestyle, reactions and interactions.

3.5.2 *Person Centered Clinical Analysis*

Knowing about the patient's local context requires, first of all, time. While parameters and values can be collected through tests and orthodox clinical enquiry, processes and interactions need to be understood through person centered dialogue. Person centered dialogue is a type of interaction in which the patient is met as a whole person: her biology, her biography, her history and her narrative are taken as equally important information for the purpose of the clinical inquiry (see also Anjum and Rocca, Chap. 4 and Low, Chap. 8, this book).

3.5.3 *Focus on Theories of Causal Mechanism*

In order to evaluate the propensity of an outcome for a single case, it is necessary to have an insight into how and why such an outcome might be generated. This requires a certain degree of general, theoretical knowledge about the process at hand. It is impossible, for instance, to evaluate the propensity of someone to develop diabetes without having an idea about the biological mechanisms underlying the onset of the illness. This requires that the clinician cultivates a high level of theoretical, pathophysiological knowledge along with statistical evidence. At the same time, as mentioned above, local causal mechanisms are of considerable relevance for the propensity approach. These include biological symptoms of broad interest that the clinician might notice in the patient, besides the symptoms of relevance for the targeted examination. But such local mechanisms also include higher level, socio-psychological mechanisms which might influence a patient's physiological conditions.

3.5.4 *Multidisciplinarity and Networking*

By adopting the propensity approach to probability, the clinician makes use of a wide number of scientific and theoretical insights, besides statistical and population studies. For instance, in order to maximise the propensity for recovery in a depressed patient, a clinician must be updated about scientific insights on the various causal mechanisms influencing the onset of depression. But the connection between research and the clinic is not a one-way street. Since the clinical search for propensities is focused on local processes and interactions, it potentially becomes a reliable source of new general scientific hypotheses about the mechanisms of healing and disease. This is particularly the case with unexpected clinical observations, such as side effects of drugs. We can illustrate this with an example.

Zolpidem is a hypnotic drug used to treat short-term insomnia. Clinicians reported a variety of anecdotal undesired and beneficial effects in Zolpidem users: sleepwalking, sleep-eating and sleep-driving, compulsory behaviours followed by amnesia, but also speech recovery after stroke, recovery of mobility after brain injury, and recovery from posttraumatic semi-unconscious state. These insightful clinical observations resulted in new hypotheses for basic research, for instance about the mechanism of recovery after brain injury. Notice that these effects are very rare and sometimes unique, therefore they would not count as particularly relevant evidence for a frequentist.

These considerations highlight that clinicians should ideally work in a multidisciplinary network with researchers, so that information can be easily shared among diverse experts.

3.5.5 *The Potential of Clinical Experience for Advancing Medical Knowledge*

We often think of the perfect medical research and healthcare system as a system that places patient care as the final aim of a long process. In a way, this is hardly controversial: patients' interests must be prioritised over commercial or other economic interests, for instance. Research hypotheses, funding, and experimental designs ought to be developed with a special consideration that they are meant to be primarily useful for the patient. Important steps are being taken in this direction, and bioethics has this as a key principle of both healthcare and research.

This conception, however, must be somehow adjusted. There is nothing "final" about the clinical meeting between practitioner and patient. Quite the contrary: each of such encounters is potentially the beginning of a new hypothesis, a challenge for established paradigms, and the springboard for broadening medical knowledge. This is not difficult to believe if we think about the history of medicine.

Many have already emphasised the value of patient centered medicine and healthcare for the final purpose of improved clinical decision making, patient care, and clinical ethics. But few have talked about the fact that a patient centered clinical approach also has a significant epistemological value: it is the best available opportunity for advancing causal knowledge in research. Expansion of knowledge does not happen in a straight line, with the patient at the end of it. It is a continuous circle of trial and success or failure, where evidence from clinical cases are looping back to pre-clinical and clinical research. The more attention that is given to the clinical cases, therefore, the more opportunities we have to improve research (Rocca 2017).

From a practical point of view, what does this entail? First of all, the clinical interview takes on a crucial role, not only for the patient's wellbeing, but also for the whole healthcare community. This important process of gathering clinical evidence should not be left to individual skills and improvisation (see Hagen, Chap. 10, this book). Medical schools should teach patient centered models of clinical communication, and should stress their key value (see Broom, Chap. 14, this book). Second, clinical evidence should be collected in databases and networked within the broad medical community (see Copeland, Chap. 6, this book). Third, researchers should recognise the primary role of patient centered evidence for the corroboration, challenge, and advance of causal knowledge.

3.5.6 What Does N = 1 Mean, Within the CauseHealth Project?

We have seen that a propensity approach to probability requires that theoretical understanding of physiology and of illness is prioritised. Statistical knowledge can sometimes be a useful tool to gain such knowledge, but it is certainly not the only type of evidence one needs in order to understand the how and why of medical phenomena. The single patient represents a major part of the causally relevant information for understanding the illness and choosing the best treatment. In CauseHealth, we sometimes summarise this central concept through the slogan *"N = 1"*, which has been a source of philosophical debate among healthcare practitioners. We give a specific meaning to $N = 1$, which is distinct from the traditional meaning of $N = 1$ trials in medical research. In the following, physiotherapist Roger Kerry provides a full explanation of the slogan's meaning:

> *"N = 1"* is a slogan used to publicise a core purpose of the CauseHealth project. $N = 1$ refers to a project which is focussed on understanding causally important variables which may exist at an individual level, but which are not necessarily represented or understood through scientific inquiry at a population level. There is an assumption that causal variables are essentially context-sensitive, and as such although population data may by symptomatic of causal association, they do not constitute causation.
>
> The project seeks to develop existing scientific methods to try and better understand individual variations. In this sense, $N = 1$ has nothing at all to do with acquiescing to "what the patient wants", or any other similar fabricated straw-man characterisations of the notion which might emerge during discussions about this notion.
>
> In Evidence Based Medicine terms, of course, an $N = 1$ trial is a randomised controlled trial involving a single subject with a random allocation of the temporal sequence of interventions. Such a trial has traditionally sat at the very top of evidential hierarchies because it offers the best scientifically controlled conditions. CauseHealth is sympathetic to such a methodology, although the clinical notion of N = 1 means much more than just this method.
>
> $N = 1$ is both an ontological claim, about the causal singularism (this means that causation is something intrinsic to the person and the situation, and does not have to be repeated in exactly the same way elsewhere to count as causation) and possibility/plausibility of the situation that each causal setting is unique.
>
> It is also a methodological claim, arguing against the idea that the individual can best be captured by searching for the relevant sub-population. Which group should represent Rani?

Women between 40 and 50 years of age? Mixed ethnic background? Norwegians? Educational status? Etc. Say we find such a group, then why assume that this group is Rani's 'twin population'? There might be all sorts of causally relevant factors that they cannot represent but that are important in Rani's individual case. $N = 1$ is about starting from the expectation that everyone is different, rather than from the assumption that everyone is statistically average. This is a fundamentally important scientific shift in how research should be operationalised and interpreted.

Despite the above, $N = 1$ thinking is not at all dismissive of population studies, and sees them as critical tools which are well suited to signalling to where causal activity may well lie. However, the above limitations of population studies related to individual clinical decision making are highlighted within the $N = 1$ notion.

Paradoxically, as we gain more data, experience, and maturity with our population research programmes, contextual analysis of such data starts to reveal that there is indeed no "one size fits all" approach to the management of much burdensome disease, for example low back pain. Such analyses are exemplars of how $N = 1$ and population data work together.

$N = 1$ is about contextualising the individual human within population data. It moves beyond a level of thinking which says "here is a patient with low back pain, let me see what evidence based interventions are available for low back pain". Rather, it is committed to understanding the human in front of us and the causally relevant factors which will influence that person's return to a desired functional level. Some of those factors will have been represented in population data, many will not have been.

Roger Kerry, 'What does CauseHealth mean by $N = 1$?', CauseHealth blog (https://causehealthblog.wordpress.com/2017/06/22)

3.6 To Sum Up…

This chapter outlined three different interpretations of the concept of probability and explained why causal dispositionalism supports an understanding of probability as propensities, and how this influences clinical decision making and medical investigations in general. For a final illustration of the difference between the three perspectives presented above, imagine a situation in which we are going to cross a bridge with a heavy truck, and we want to evaluate the probability that the bridge will endure the weight of the truck (and consequently the risk of an accident). The frequentist approach would face this challenge by looking at how often similar bridges collapsed under the weight of similar trucks. The Bayesian approach would treat the probability as a subjective matter that changes depending on the information we have about the bridge and the truck, and would treat the probability as the value of how certain we are that an accident will (or will not) happen. The measure of such certainty will be updated every time we gain a new piece of information. The propensity approach would describe the probability using the qualities of the bridge, the truck, and the whole situation, and trying to understand the intrinsic disposition of the bridge to collapse under a certain weight. Such intrinsicality will be evaluated by investigating the properties at hand (height, length, solidity, material) and by understanding the causal and physical processes involved. All these perspectives – frequencies, uncertainty and propensities – offer something that can be useful for expanding our causal knowledge. The philosophical question is which we take to be basic.

References and Further Readings

Anjum RL, Mumford S (2018) Causation in science and the methods of scientific discovery. Oxford University Press, Oxford
Anjum RL, Mumford S, Myrstad JA (2018) Conditional probability from an ontological point of view. In: Anjum RL, Mumford S (eds) What tends to be. The philosophy of dispositional modality. Routledge, London, pp 101–114
Bohm D (1957) Causality and chance in modern physics. Routledge, London
Cartwright N, Hardie J (2012) Evidence-based policy. A guide to do it better. Oxford University Press, Oxford
Clarke B, Gillies D, Illari P, Russo F, Williamson J (2013) The evidence that evidence-based medicine omits. Prev Med 57:745–747
Clarke B, Gillies D, Illari P, Russo F, Williamson J (2014) Mechanisms and the evidence hierarchy. Topoi 33:339–360
Gillies D (2000) Varieties of propensity. Br J Philos Sci 51:807–835
Gillies D (2018) Causation, probability, and medicine. Oxford University Press, Oxford
Mellor DH (1971) The matter of chance. Cambridge University Press, London
Popper K (1959) The propensity interpretation of probability. Br J Philos Sci 10:25–42
Popper K (1990) A world of propensities. Thoemmes, Bristol
Rocca E (2017) Bridging the boundaries between scientists and clinicians. Mechanistic hypotheses and patient stories in risk assessment of drugs. J Eval Clin Pract 23:114–120
Rothwell PM (2005) External validity of randomised controlled trials: "to whom do the results of this trial apply?". Lancet 365:82–93
Rothwell PM (2006) Factors that can affect the external validity of randomised controlled trials. PLoS Clin Trials. https://doi.org/10.1371/journal.pctr.0010009
Wallmann C, Williamson J (2017) Four approaches to the reference class problem. In: Hofer-Szabó G, Wroński L (eds) Making it formally explicit, European studies in philosophy of science, vol 6. Springer, Dordrecht
Williamson J (2010) In defence of objective Bayesianism. Oxford University Press, Oxford

Open Access This chapter is licensed under the terms of the Creative Commons Attribution 4.0 International License (http://creativecommons.org/licenses/by/4.0/), which permits use, sharing, adaptation, distribution and reproduction in any medium or format, as long as you give appropriate credit to the original author(s) and the source, provide a link to the Creative Commons license and indicate if changes were made.

The images or other third party material in this chapter are included in the chapter's Creative Commons license, unless indicated otherwise in a credit line to the material. If material is not included in the chapter's Creative Commons license and your intended use is not permitted by statutory regulation or exceeds the permitted use, you will need to obtain permission directly from the copyright holder.

Chapter 4
When a Cause Cannot Be Found

Rani Lill Anjum and Elena Rocca

> In Western countries, most persons asking their regular general practitioner for help and advice share certain common characteristics: they show up repeatedly and over time, although at varying intervals and for a variety of reasons; they present complex health problems which may involve acute maladies but often include chronic, somatic and/or psychiatric distress; at the same time, they may seek advice for medically unexplained or undefined malfunctions, which may be equally if not more problematic and incapacitating than the supposedly well-defined diseases or disorders…
>
> Experienced GPs are aware of patterns of sickness, both within groups of patients and in individuals, that seem to point to sources of bad health beyond the medically defined horizon of causality. These patterns are complex and transgress such medical dichotomies as "somatic" and "mental". They are specific in the sense that they represent clusters of diseases or malfunctions, which apparently have such common "causes" as inflammation, infection or invasion (in the sense of tumour growth). This, however, leads to the next level of relevant questions, those regarding the "cause" or "causes" of a dangerously compromised immune system manifesting in systemic inflammations, repeated infections or multiple invasive processes. Here, a rapidly growing documentation highlights the medical significance of context, offering ways of understanding the detrimental impact of lifetime adversity on health.
>
> Anna Luise Kirkengen, 'Map versus terrain?', CauseHealth blog (https://causehealthblog.wordpress.com/2017/04/18)

4.1 The Clinical Challenge of Medically Unexplained Symptoms (MUS)

Healthcare professionals are regularly faced with patients who suffer from multiple conditions at the same time. How exactly these conditions relate is not a straightforward question and, in some cases, the causes themselves remain a mystery. Patients who experience what are commonly referred to as *medically unexplained*

R. L. Anjum (✉) · E. Rocca
NMBU Centre for Applied Philosophy of Science, Norwegian University of Life Sciences, Ås, Norway
e-mail: rani.anjum@nmbu.no; elena.rocca@nmbu.no

© The Author(s) 2020
R. L. Anjum et al. (eds.), *Rethinking Causality, Complexity and Evidence for the Unique Patient*, https://doi.org/10.1007/978-3-030-41239-5_4

symptoms, or MUS, exhibit a number of symptoms that appear together but do not seem to have a single, common biomedical cause.

The increase in medically unexplained symptoms represents an emerging problem in European and other industrialised countries. 'Medically unexplained' refers to the lack of explanatory pathology. Researchers have not been able to find a common set of causes, a definite psyche-soma division, or even clear-cut classifications for these symptoms. The problem with these conditions not being explained generally means that the biological causes of the symptoms are unknown. Some of the causal factors involved might be known, but the underlying mechanisms are not understood. In general, no adequate psychological or organic pathology can be found, and medical examination is unsuccessful in giving a diagnosis to the symptoms (Eriksen et al. 2013a). Each patient seems to have a unique combination of symptoms and a unique expression of the condition, and medical uniqueness appears to be the rule rather than the exception. A problem with this is that evidence from population studies are of limited use for these patients.

That no causal explanation is found for a condition is not a rare or unfamiliar phenomenon. MUS have been estimated to account for up to 45% of all general practice consultations, and a study from secondary care suggests that after 3 months, half of the patients received no clear diagnosis (Chew-Graham et al. 2017). It is difficult to give a precise number, however, since there is no general agreement over what counts as a MUS. Some examples are chronic fatigue syndrome, irritable bowel syndrome, low back pain, multiple chemical sensitivity, general anxiety disorder, tension-type headache, post-traumatic stress disorder and fibromyalgia. Some other conditions lack a commonly accepted diagnosis, or even a clear definition, yet they seem to be increasingly common, in some cases almost mass phenomena. Another example of a medically unexplained condition is *burnout*, which indicates a pathological condition somehow connected with severe stress and work overload. Psychotherapist Karin Mohn Engebretsen has dedicated her doctoral research to the analysis of burnout as a challenge for the current scientific paradigm. She writes:

> As a Gestalt psychotherapist, I have seen an increasing number of individuals over the last fifteen years that experience themselves as burned out. This fact has triggered my interest to explore the phenomenon further. Burnout is a medically unexplained syndrome (MUS). As with other MUS, there is a tendency to assume a narrow perspective to focus on problems related to psyche or soma as pathologies located exclusively within the patient. Research has mainly looked for clear-cut one-to-one relations between cause and effect. These relationships are however difficult to find in complex syndromes.
>
> Burnout might instead be seen as a reaction to complex causes and a broad contextual setup, but unfortunately, such point of view has only been marginal. Consequently, medical professionals are faced with comprehensive challenges due to factors such as lack of a causal explanation, lack of diagnostic descriptions and lack of a treatment or medical interventions.
>
> Karin Mohn Engebretsen, 'Are we satisfied with treating the mere symptoms of medically unexplained syndromes?', CauseHealth blog (https://causehealthblog.wordpress.com/2017/03/27/are-we-satisfied-with-treating-the-mere-symptoms-of-medically-unexplained-syndromes/)

As pointed out by Engebretsen, healthcare professionals who deal with a person experiencing burnout will face some deep theoretical and methodological issues (Engebretsen 2018; Engebretsen and Bjorbækmo 2019). For instance, is a response to stress overload to be considered a medical condition? Should it be seen as one of the symptoms, or one of the causes? How to distinguish burnout from other well-defined pathologies with similar symptoms, such as depression? And, even more problematically: how to act when no clear-cut causality can be found, given that curing a disease means to counteract its causes?

The problem of understanding MUS could be interpreted as an empirical matter, to be solved by doing more of the same. On this view, more observation data, RCTs, symptom measurements and classification could ultimately lead to a clearer understanding of these conditions. Given the dispositionalist framework of CauseHealth, however, we see MUS as a symptom of some deeper problems in current medical thinking (Eriksen et al. 2013b). Specifically, the problem of MUS seems to point to a philosophical challenge, namely: *how to understand causality in cases of complexity, individual variation and uniqueness.*

The challenge of dealing with MUS, even conceptually, played a central role in the CauseHealth project. We started from the idea that the problem of MUS is a practical challenge for medicine, but one that has a philosophical source. MUS are troubling and chronic conditions that are often depicted as outliers: atypical illnesses where standard causal explanation fails. From the dispositionalist perspective, however, every patient is to be considered, in one way or another, an outlier. Recall that dispositionalism is a singularist theory of causality (see Anjum, Chap. 2, this book). Since causality happens in the single case, we need to be armed with strategies to look for it in the single patient, while at the same time making use of general medical and other theoretical knowledge. The problem of dealing with MUS, then, is rooted in a deep conceptual challenge of the current paradigm. Finding a way to deal with these conditions epistemologically was therefore seen as the key to getting a better grasp of medicine as a whole. If we understand the problem of MUS, we thought, we will better understand the problem of investigating causes of health and illness generally.

In this chapter, we take a closer look at the challenge that causal uniqueness represents, not only for the healthcare professional having to deal with MUS and other complex conditions, but also for the whole medical paradigm. We make a 'philosophical diagnosis' of the problems of dealing with causal uniqueness in the clinical encounter: they come from a positivist, or Humean, understanding of causality. We then explain how an ontological turn toward a dispositionalist starting point should help us deal better with the challenging features of MUS: causal complexity, heterogeneity and medical uniqueness. From a dispositionalist perspective, we will argue, these features should not be seen as problems for causality, but instead as *typical* for it, and therefore as opportunities to understand causality better.

4.2 The Problem of Uniqueness

While in medicine and healthcare the default assumption is that all patients are different, causation itself is sought as something that is robust throughout different contexts. This leaves us, in effect, to search for same cause and same effect:

- Same symptoms, same diagnosis (diagnostics)
- Same diagnosis, same intervention (standardised treatment)
- Same intervention, same effect (tested though RCTs)

Although individual variations are acknowledged, they are nevertheless not the focus when trying to establish causality. Instead, variations can be used to form more fine-grained classifications or sub-groups, where one again looks at what is the same. In other words, uniqueness is considered an obstacle when one tries to establish causality scientifically.

This contrasts with the dispositionalist framework. No two individuals will have exactly the same combination of causal dispositions or propensities. Even if there are some dispositions that we share, such as gender, age or medical condition, so many other dispositions will be different from one individual to another. Grouping patients into more relevant sub-groups will plausibly tend to give a more appropriate average than broader and unspecific sub-groups. We know, however, that not all pregnant women in their thirties or all men over 60 with hypertension are identical in all their dispositions – or even sufficiently identical. Which of these dispositions are taking part in the single causal process that we are investigating? This is a question that cannot plausibly be answered with certainty. We will therefore never be sure of whether or how precisely a sub-population represents the dispositions in place in the individual process. A frequentist approach, we have said, will either have to overlook this knowledge gap or try to further specify the relevant sub-group (see Rocca, Chap. 3, this book). Eventually, however, one might end up with a sub-group with only one member: the $N = 1$ group consisting of the single patient. Still, the problem remains how to establish, predict and explain causality for a patient for whom no suitable, or suitable enough, sub-population can be found.

This is one reason why MUS represent a methodological challenge for medicine and healthcare. In the current paradigm, the best way to establish causality is by showing that the *same* cause makes a difference toward the *same* effect in sufficiently similar contexts. To make this clear, think back to the principle of randomised controlled studies (RCTs), as explained above in Sect. 3.2. We saw here that RCTs are considered to be the best way to establish causality within the current paradigm of evidence based medicine and practice, and they are designed to test for a type of homogeneity: common causes and common effects. This means that even though there is plenty of individual variation within the clinical study, these variations are not what the RCT is designed to study or establish. On the contrary, such individual variations are supposed to be shielded off through randomisation, so that test group and control group are *overall* very similar. With RCTs, we look for the overall effect of an intervention in the test group, compared with the overall effect in the control group. The intervention is then the same, and the effect tested is the same.

RCTs thus target *same cause* (intervention), *same effect* (outcome). This is completely in line with Hume's regularity theory of causality (Hume 1739), but it doesn't acknowledge the dispositionalist perspective that causes as dispositions are *intrinsic* properties: they *tend to* manifest, but not always. They *tend to* make a statistical difference, but not always. They *tend to* produce one effect, but not always the same (Anjum and Mumford 2018, see also Anjum, Chap. 2, this book). RCTs are great tools to detect manifestations that make a difference at population level, but they are not useful for studying dispositions that remain mostly unmanifested, and which tend to manifest themselves in single and causally unique cases. We see, then, that the problem of MUS is not an isolated one, but one that has its roots in the Humean influence on medical thinking about causality that can be summarised in the following three points:

1. A and B are observed repeatedly (empiricist criterion: causality must be detected empirically)
2. Whenever A, B, under some normal or ideal conditions (regularity criterion: same cause, same effect)
3. B happens because of A (monocausality criterion: one cause, one effect)
4. If not B, then not A (falsification criterion: a difference in effect must mean that there is a difference in the cause).

For A to be the cause of B, these conditions must be met, according to the Humean notion of causality. Medically unexplained symptoms, however, typically fail to meet one or several of these criteria, which is why we cannot say that a medical cause has been found.

Let us show this by considering the case of unspecific lower back pain. Qualitative studies show that in the clinical dialogue, patients usually associate this condition with an episode such as bending or lifting. Patients mention that they felt sudden pain during a certain activity, and that they have been in pain ever since (Jeffrey and Foster 2012). There is, in the clinical encounter, a deep intuition of a causal link between a certain accident, or event, and the condition. However, this cannot be epidemiologically confirmed. There is much literature on unspecific lower back pain, but no systematic association has been shown with mechanical factors (lifting, standing, walking, postures, bending, twisting, carrying, and manual handling) nor with activity levels, obesity, smoking, mood, or genetic factors (see Eriksen et al. 2013b for a review of the epidemiological evidence). None of these causal factors seem to fulfil the Humean criteria of regularity, repeatability and falsification. Epidemiologically, and according to Humean criteria, therefore, there is no clear cause of unspecific lower back pain, despite decades of research. And yet, single patients tend to be able to indicate a cause, at least as they experience it. This and similar cases of medically unexplained conditions represent a challenge for any attempt at standardisation or universal approach to cure and healthcare.

4.2.1 *The Patient Context: What Was There Before*

From a healthcare perspective, one does not expect that the same cause will give the same effect in different individuals. Individual responses depend on what else was there already, as part of the patient's own context. A person who is at a vulnerable stage in life might be more disposed to an infection than a person who is at a more robust stage, for instance. This is, one can say, elementary clinical knowledge. Still, this real-life complexity becomes a problem for causal understanding when we try to analyse it using the Humean criteria. Dispositionalism instead acknowledges complexity and context-sensitivity as basic features of causality.

We can represent the different impact of one causal disposition in different patients with the vector model of causality (Mumford and Anjum 2011), where each vector represents one causal disposition in place, and the line T represents the threshold for the manifestation of an effect, as explained in Chap. 2. In the vulnerable patient (Fig. 4.1), the situation is much closer to the threshold of illness than in the robust case (Fig. 4.2). This means that even a minor burden on health can have a major impact, because it pushes the situation over a threshold. This is a well-known phenomenon. We often speak of the straw that broke the camel's back, which was simply the final straw adding to the already heavy burden. A cause might then be simply what tips the situation over the threshold, which seems far too insignificant if we ignore what was already there before it.

Let us say that the two patients get affected by influenza, and after that only one of them develops a chronic burnout syndrome, while the other recovers normally.

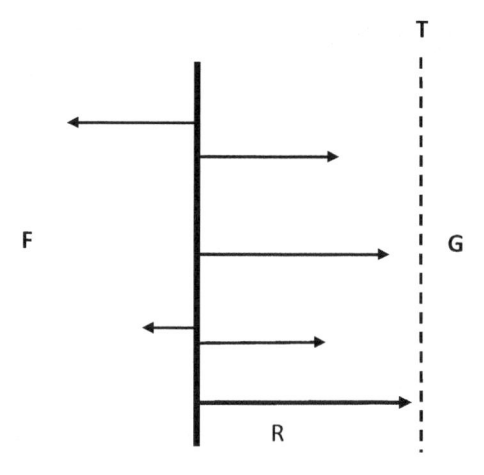

Fig. 4.1 A vulnerable situation, where R is close to the threshold for illness

Fig. 4.2 A robust situation, where R is far from the threshold for illness

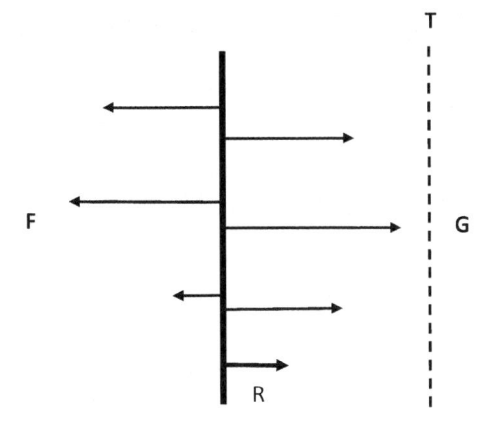

From a Humean point of view, this does not tell us much about the causal role of the influenza for the onset of burnout. Instead, we would need to check whether there are other patients as similar as possible to the patient who develops burnout symptoms after getting influenza (same cause – same effect, *all else being equal*).

From a dispositionalist perspective, however, looking further into cases of individual variation and context-sensitivity represents a chance for understanding something about the underlying causal story. When the same cause gives different effects in two different contexts, we might learn something new. Clearly, to do that, it does not help to focus on the single cause or the single effect. Instead, one should try to understand what was already there in the two different contexts, disposing toward or away from health and illness. This type of reasoning needs to be *qualitative* and *explanatory*, in order to be fruitful. By trying to understand all the causal dispositions in place, and the way they interact with each other, we build a causal explanation – a hypothesis – for how and why things went the way they did.

Note that the causal explanation, or causal mechanism, although being based on empirical evidence, is not itself something we can observe directly. This is why Hume's empiricist approach does not include causal theories or explanations but sticks to what can be observed and counted. On the contrary, the search for evidence of a plausible causal explanation (which dispositions are present, how they interact and manifest) is at the core of the dispositional approach. It is also crucial for any scientific theory, including in medicine and healthcare.

Humean and empiricist influence have been strong, not only in research, but also in the clinic. There tends to be the expectation, at least in the implementation of some health policy and clinical guidelines, that patients with the same diagnosis should respond similarly to the same treatment. Personalised medicine and system medicine have been rising trends and can be seen as attempts toward a more dispositionalist approach: aiming to fit the treatment to the patient's own dispositions. However, these approaches are mainly focussed on genetic or molecular dispositions and have less focus on psycho-social or ecological complexity (Vogt et al. 2014). This will be discussed in Chap. 5 when we look at the biomedical model of medicine.

Allowing the features of uniqueness and complexity to guide the clinical encounter, we should focus less on what is the same and more on what is different and unique for this particular patient, also for causal matters.

> Simply put...
>
> Humeanism refers to David Hume's regularity theory of causality, which emphasises features such as empiricism, observable features (data), monocausality, repetition and same cause – same effect. Probabilities are understood as generated statistically (frequentism).

4.2.2 Qualitative and Quantitative Approaches to Causal Inquiry

What exactly do we mean by a qualitative analysis? And how should such analysis help us look for causal explanations? How does this approach contrast with quantitative analysis and the search for same cause – same effect? In our philosophical framework, we have a particular take on what should count as qualitative and quantitative approaches to causal inquiry. Qualitative approaches will be concerned with the investigation of many types of information in few tokens, and with how these relate, and under which conditions. In contrast, quantitative methods will look for few types of information that are in common for many tokens (see also Anjum and Mumford 2018: 106). Notice that from the dispositionalist perspective, qualitative research can advance causal understanding and theory, and is not limited to the purposes of meaning and lived experience (see Sect. 4.4). Qualitative research, in our definition, encompasses scientific enquiry of a phenomenon, as long as such enquiry aims to understand a causal process, while quantitative research aims to identify the numerical relationship between variables. A qualitative approach, in our understanding, might involve numerical values, but is always process-oriented, aims to generate theoretical understanding, it is adapted to the most relevant context of application of the research, and it happens in-situ, often in a participatory way.

An example might help illustrate this distinction.

> For example, a recent large study compared the whole-genome sequences of participants with food allergy to peanuts, egg or milk with non-allergic participants (in total almost 3.000 individuals were included)[...] The results showed statistically significant DNA modifications in specific loci of the genome, indicating that these loci are probably part of the genetic component of the food allergy. Other information about the participants were age, sex, ancestry (European or non-European), results of food allergy tests, and presence of other allergy-related disease. While such a horizontal analysis has big statistical power, it relies on the preliminary selection of a limited amount of variables to compare. The selection is informed by existing knowledge and working hypothesis (in the case of this study, that allergies have a genetic component). Additionally, it is dictated by practical considerations since these studies include a large amount of participants. While results are statisti-

cally robust, their contribution is limited to a small part of the picture. In fact, genetic predisposition is only one of many actors for the onset of a condition.

Let us imagine using a larger filter to evaluate which contextual variables to include in the analysis. We might then consider a complete range of clinical factors, blood levels, present and former state of health, dietary habits, lifestyle, polypharmacy, psychological health, addictions, traumas, including as much information as possible about the unique context that was exposed to the allergen. This would necessarily restrict the comparison to a limited number n of patients […].

The experiment would then be a qualitative, rather than quantitative, analysis. It would have a different aim: the aim of identifying not a single element that is frequently involved, but enough elements to suggest a pattern, or offer an explanation that is valid in this specific instance. Such explanation might fall outside the boundaries of existing knowledge and suggest an advance in the overall understanding of causal mechanisms. Finding out whether these hypotheses are generalizable and to which extent, belongs to a subsequent stage of research. (Rocca 2017: 117)

> Simply put…
>
> In our framework we propose that qualitative approaches to causal inquiry collect many types of information in few tokens and look for a theoretical understanding of how these relate causally in a particular context. Quantitative approaches, instead, look for few types of information in many tokens, and aim to identify numerical correlations among them.

4.2.3 Dispositional Take On Perfect Regularity: Is It Causality or Something Else Entirely?

A dispositionalist denies that causality is something that produces perfect regularity between cause and effect. Instead, causality is understood as tendencies, where the cause A only tends or disposes toward the effect B. So even if A is present, B can still be counteracted by adding an interferer *I* (Mumford and Anjum 2011). All of medicine is premised on this idea. Even if one has not yet been able to find a treatment, the expectation is still that if one can understand the causal mechanisms of the disease, it should be possible to counteract or interfere with the causal process in one way or another.

This has a surprising consequence. If there were to be a perfect correlation of A and B, where no changes in context could influence the situation in any way, a dispositionalist should become suspicious. Is this a case of causality after all? Or could it be a case of classification or identity? For instance, all humans are mortal. And although scientists are still working on ways to counteract and delay death, one could still argue that any immortal being could not be human. So even if there were a human-like immortal species in the future, we might say it's not the same as a human.

Now take a medical case. It is said that Down syndrome is a genetic disorder caused by the presence of all or part of a third copy of chromosome 21. But is this the right way to phrase it? As long as a person has the extra chromosome, they will be diagnosed with Down syndrome. If Down syndrome is then *defined* as the condition of having the third copy of chromosome 21, then, of course, there will be a perfect regularity between A and B. But the reason why we have the situation that whenever A then B, is that A is defined as B. In that case, A = B. This does not mean that there is no causality going on here. The causal relationship would then be between the extra chromosome and the expression of the condition, which vary in degree from individual to individual. The *symptoms* of Down syndrome are then caused by the extra chromosome, and will be manifested in different ways in different individuals. *Whether* someone has the syndrome will correlate perfectly with whether they have the extra chromosome, without any individual variations. This suggests that it is an identity relation, not a causal one.

Perfect regularities, on the dispositionalist perspective, could be produced by other types of truths than causal ones, such as classification (all humans are mammals), stipulation (all electrons are negatively charged), identity (bachelors are unmarried men) or essence (humans are mortal). In contrast to these types of claims that have the categorical form 'All As are Bs', causal claims are about what happens under certain conditions. A causal claim is therefore a hypothetical or conditional matter: 'If we do x to y, would z follow?'. To say that 'All As are Bs' is to make a statement about how to categorise A and B with respect to one another. In contrast, when we ask whether A is a cause of B, we want to know whether and to what degree A is able to bring about B or at least contribute to the production of B.

Note, however, that if A is indeed a cause of B, a dispositionalist should not expect that all instances of A will actually and successfully produce B. Causes, as dispositions, are irreducibly tendential. There is never more than a tendency of A to produce B. As discussed in Chap. 2, a dispositional tendency can be stronger or weaker. Someone can be more or less vulnerable, more or less violent and more or less allergic to peanuts, for instance. We also saw that dispositional tendencies give rise to individual propensities, rather than statistical frequencies (see Rocca, Chap. 3, this book). The degree of tendency does not determine how often a disposition will manifest, but only how strong the intrinsic disposition is in this individual situation. For instance, if we want to know how fertile someone is, one should do a sperm count rather than counting the number of offspring. The higher the sperm count, the stronger the disposition of fertility. It does not follow from the strong fertility that one will eventually have a lot of children. It also does not mean that other people with the same sperm count will have many children.

We see, then, that a dispositionalist should not expect perfect regularity of cause and effect. Instead, a dispositionalist should be sceptical if there is a perfect correlation that is insensitive to contextual change. Could it be a case of identity, classification or essence instead? Or have we already stipulated some ideal conditions or idealised model under which the cause would always produce the effect? Either way, we cannot expect that causality will manifest itself in perfect correlation in a real-life situation such as what we encounter in the clinic. The only way we can expect that the same cause will always produce the same effect, is by stipulating some average, normal or ideal patient with average, normal or ideal responses. In the clinic, however, such encounters are rare.

4.3 An Important Lesson from Medically Unexplained Symptoms (MUS)

We saw that medically unexplained symptoms remain a challenge for the healthcare profession because of some problematic features: causal complexity, heterogeneity and medical uniqueness. If this is a problem for establishing causality, then we got a much bigger problem than MUS. In most health conditions, there is at least some complexity of causes, some individual variation and some unique factors. This is the case for cancer, heart disease, obesity, Alzheimer's, hypertension, diabetes, stress-related symptoms and many other conditions. Although these conditions are not medically unexplained, because there is some common pathology, they are still complex disorders with multiple causes.

All conditions that are caused by a combination of genetic, environmental and lifestyle factors, will have many unknown causes and many causes that are unique to that patient (Craig et al. 2008). And since each patient has a unique combination of biological, social and psychological factors, complex diseases are very likely to be heterogeneous (Hellhammer & Hellhammer 2008). From the dispositionalist perspective, therefore, there is a lot more in common between medically unexplained and medically explained conditions than what is normally assumed. What can we learn from this?

First of all, this means that our understanding of illness cannot rely solely on single or few physical, or biomedical, homogeneous causes. When a common physical cause of illness is found, such as the bacterium helicobacter pylori (HP) for peptic ulcer, it can quickly become the main focus of medical attention while other causes gain the status of 'background conditions'. This is the case with lifestyle factors such as stress and diet, for instance, which were thought to cause ulcer before the discovery of HP, and have been since decades at the periphery of the therapeutic focus (de Boer and Tytgat 2000). Looking at this case in more detail, however, it has been estimated that at least half the world's population is infected by the HP bacterium, but most of those infected never develop an ulcer (Go 2002). We see then that although one causal factor might be the necessary condition for the development of a pathology, whether such pathology is triggered, how it is expressed, when, and to what degree, will be influenced by a plethora of other causal factors. When dispositionalism emphasises causal complexity, it means that a mono-causal focus on a common physical cause will necessarily mean that we miss out on some of the causal story, if not most of it (Copeland 2017). This is why CauseHealth proposes that the challenging features of MUS should be treated as the norm, rather than be dismissed as marginal and atypical. All illness is complex, and many of the causal factors will be unique to the individual case.

This has a practical consequence, not only for research, but also for the clinical encounter. It means that by understanding the complexity and uniqueness of the patient's situation, one will find a number of factors that might be influencing their condition positively or negatively. And although the known biomedical factor (such as the HP bacterium) leaves little wiggle-room other than the standard medical interventions, many of those other influences (lifestyle, diet, stress) are usually important to target too, and might be even easier to counteract. Being aware of these other dispositions that are causally relevant for the health condition, and understanding how exactly they influence the experience of health and illness, can then empower both patient and clinician. The patient might get a better understanding of what caused their condition, but that is not all. The patient might also be able to influence and work with some of these dispositions, thus getting a better sense of control and agency with respect to their own health (see Price, Chap. 7, this book).

4.3.1 We Need Many Methods to Establish Causality

Searching for and establishing causes in the dispositionalist framework is not something that can be done using only one single method, such as RCTs (Rocca and Anjum 2020a). In CauseHealth, we have argued that causal enquiry requires a plurality of methods, each of which picks out one or more symptoms of causality (for details, see Anjum and Mumford 2018). In an open letter to *BMJ Evidence Based Medicine*, co-signed by 42 clinicians and philosophers from international and interdisciplinary research networks working specifically on causality in medicine, we urged that EBM approaches widen their notion of causal evidence:

> The rapid dominance of evidence based medicine has sparked a philosophical debate concerning the concept of evidence. We urge that evidence based medicine, if it is to be practised in accordance with its own mandate, should also acknowledge the importance of understanding causal mechanisms… Our research has developed out of a conviction that philosophical analysis ought to have a direct impact on the practice of medicine. In particular, if we are to understand what is meant by 'evidence', what is the 'best available evidence' and how to apply it in the context of medicine, we need to tackle the problem of causality head on… In practice, this means understanding the context in which evidence is obtained, as well as how the evidence might be interpreted and applied when making practical clinical decisions… It also means being explicit about what kind of causal knowledge can be gained through various research methods. The possibility that mechanistic and other types of evidence can be used to add value or initiate a causal claim should not be ignored. (Anjum et al. 2018: 6)

How exactly would this work if we assume a dispositionalist understanding of causality?

In dispositionalist terms, if we want to establish a causal link between A and B, this corresponds to establishing whether A has intrinsic dispositions that, in combination with other dispositions, can eventually produce B. Different methods will have some strengths and some limitations for the purpose of finding such intrinsic dispositions. For instance, RCTs are good for picking out which factors make a difference, or raise the probability of an effect, on population level. But positive results from RCTs do not guarantee that this difference-making or probability-raising happened *because of* some intrinsic disposition in the test group, since that would require a further theoretical explanation. RCTs are also less suitable for testing the contextual complexity of mutual manifestation partners, since the focus is on one or few particular interventions (for which there can be a control or comparison) and one or few outcomes. Other methods could be more suitable than RCTs when we are searching for dispositions or manifestations that are too rare to show up in statistical approaches. In this case, retrospective case-control studies allow us to study outlier cases or very rare conditions. However, since these studies are designed to find common dispositions for the same outcome across different contexts, any non-common dispositions contributing to the outcome in the individual cases will not be targeted.

These are just two examples, but all scientific methods will similarly be designed to test some specific symptoms of causality, while other symptoms will fall outside the scope of the test. The problem is if we think that one method should be a perfect test for picking out causality. Such a perfect test might require that we *operationalise* causality. This means that we simply identify the phenomenon of causality with the method we use to test it. Examples of operationalization can be to identify temperature with the measure shown on the thermometer, depression with a series of specific behaviours, or cancer with a positive screening. In the case of causality, operationalisation might correspond to saying that causality is nothing more than the statistical difference of effect between experimental group and control group, as detected for instance by a positive RCT.

Operationalisation of causality would be a perfectly acceptable strategy for a strict empiricist, since they would reject any ontological reality that cannot be observed. Indeed, Hume already stated which observable features would be necessary and sufficient for calling something causality: constant conjunction, temporal priority and contingency. From a dispositionalist perspective, however, there would be no one perfect test of causality that could empirically pick out all its features. A cause will tend to make a difference, but there is no perfect overlap between difference-making and causality. A cause will also tend to produce some regularity, but again there is no perfect overlap (Anjum and Mumford 2018).

Embracing the idea of methodological pluralism – that we need more than one method to establish causality – we will now look at another method for obtaining more qualitatively rich causal information, namely patient narratives. We make a case for the epistemological importance of obtaining detailed information from the patient herself when searching for causal explanations of illness. As a methodology, rich patient information allows the detection of relevant dispositions that are uniquely combined in this individual patient (biomedical, biographical, lifestyle, and life situation), and that could be relevant for their condition or the treatment. This type of knowledge should therefore not be ignored if we understand causality in a dispositionalist sense.

4.4 Patient Narratives as a Way Forward

We have said that the clinical encounter can contribute to improving the causal understanding of health and illness, in the dispositionalist sense. We saw that, in order to evaluate individual propensities, one needs to learn as much as possible about the dispositions and interactions in place in the particular case of interest (see Rocca, Chap. 3, this book). In the case of healthcare, this means learning more about the patient and their context. From the dispositionalist perspective, therefore, the clinical encounter has a pivotal role to play for advancing our understanding of the causal story behind illness and suffering. Notice that we do not talk only about the clinical examination, which is just a part of the encounter. In the *examination*, the clinician collects biological and medical information, while in the *encounter* she meets the patient as a person with a unique biography, context and story. Osteopath Stephen Tyreman argues that a whole person centred approach is crucial for understanding illness and symptoms.

> What do symptoms tell us about the *person* rather than their disease? Symptoms are key elements in a person's narrative about illness in general and their illness in particular. We want to know what symptoms mean, what they tell us (in a narrative sense) about what has happened and what the future will be like. As much as indicating a particular biological problem, symptoms reflect how we live—the smoker's cough, the athlete's muscle ache, the workaholic's tiredness, the sedentary person's breathlessness, and so on. Are these indicators of actual or potential disease or is disease a possible emergent outcome of such behaviour? Many symptoms we accept as normal and healthy—the discomfort of pregnancy and childbirth or the stiffness after a day's physical activity, for example. In other words, do symptoms tell us more about a person's living before they tell us about disease?
>
> Stephen Tyreman, 'More on symptoms', CauseHealth blog https://causehealthblog.wordpress.com/2017/04/03/more-on-symptoms/

How exactly can the clinical encounter contribute to improving causal knowledge of the dispositions in place? The way modern medicine is generally practiced does not seem to leave much space for understanding the full complexity of the causal situation of patients. In the last century, for instance, the objectivity of doctors' reports was emphasised. Physicians have to 'translate' patient reports and accounts of their condition into a standard medical language. This is not in itself a limitation. The problem is when such standard medical language is seen as the only information of significance, while the information that is excluded, about the patient's version and interpretation of their condition, is considered irrelevant for the purpose of diagnosis and treatment (for an example of this, see Kirkengen, Chap. 15, this book).

There is an alternative to this orthodox practice that is more in line with the dispositionalist framework. This is to use patient narratives as an essential part of the evidence available (Greenhalgh and Hurwitz 1998). We see this in a recent development in medical humanities, called narrative medicine, although it would not generally be considered as *causal* evidence. Instead, patient narratives are often dismissed as causally irrelevant because of their anecdotal nature, a story from a single unique patient. In CauseHealth we have promoted the importance of qualitatively rich information about the dispositions provided by the local context (the patient), based on causal singularism (medical uniqueness), context-sensitivity (heterogeneity) and genuine causal complexity (holism and emergence) (Rocca and Anjum 2020b). The main idea that we here want to emphasise is that the subjective and embodied experience as told by the patient, traditionally seen as a possible obstacle to the objective medical diagnosis, is in fact a powerful tool for exactly that purpose. Patients, in their narrations, choose to include some information and to omit some other. A narration also gives meaning to the medical events, and such meanings will most probably vary depending on the narrator. Crucially, there is a reason for such interpretations and this reason is not possible to capture in the standard medical story.

We can illustrate this point with an example from the treatment of morbid obesity. Guidelines recommendations for this condition are based on knowledge about the biological mechanisms underlying normal and irregular food intake and appetite. Recommended treatments consist in lifestyle modification programs, pharmacotherapy and bariatric surgery. Depending on the single patient and her standard medical story, physicians can choose the treatment that is thought to be best suited for the patient. Although all these interventions could result in some modest to good improvement, there is a systemic tendency toward re-gain of weight in the vast majority of the patients, provoking frustration in both clinicians and patients (Karmali et al. 2013).

What might happen if the standard medical story is accompanied by a thorough analysis of patient narratives? One answer comes from general practitioner Kai Brynjar Hagen. He has been working for many years as a senior consultant at a Norwegian regional centre for morbid obesity. His job is to assess the patients before bariatric surgery, which is the last step in the therapy of obesity. The assessment included interview sessions with patients. During these sessions, Hagen collected the stories of their embodied, lived illness experience. He came to the hypothesis that what practitioners treat at the moment (energy homeostasis imbalance, food intake, lifestyle) are in the majority of cases only the symptoms or the condition, and not what he understands as the real or core cause of obesity. By taking the time to listen to his patients, asking them about their childhood, whether they enjoyed school, how their family life is and how they feel about their life in general, he observed that many of his patients had experienced some sort of emotional traumas that affected, triggered or worsened their eating disorder. His worry is that if these biographical aspects are ignored and obesity is treated as a purely biomedical condition, to be solved by eating less, one will fail to target the true cause of the problem (see Hagen, Chap. 10, this book). This could also explain why the current treatments for obesity in many cases fail and why there is a tendency toward suicide among patients who have undergone bariatric surgery (Lagerros et al. 2017; Neovius et al. 2018; Castaneda et al. 2019).

When overlooking important causes of obesity, such as the dispositions of trauma that manifest themselves in an eating disorder, one fails to target the source of the problem or even the core cause. If a person is obese but used to be anorexic or bulimic, then the solution to overeating is not to go on a diet. That might trigger a relapse into the other extreme, of undereating. This shows the importance of understanding the whole causal story of complex conditions, of which obesity is just one example. But these causes cannot be identified without taking the narrative of each single patient as an essential part of the available causal evidence.

4.5 Using Patient Narratives

Patient narratives can be different things and we will here mention three examples. They are all from the clinical encounter but represent slightly different approaches (see also Solomon 2015).

4.5.1 Narrative as a Tool for Causality Assessment

When a new drug is introduced on the market, it must be monitored for risk and safety purposes. In this process, patient narratives about possible side effects of medications are collected systematically and used as the basis for causality assessment and hypothesis generation. Some of these narratives come from the patients directly and into global databases such as *VigiBase*. Other narratives come from clinicians or from pharmacists, but in those cases, the narrative is interpreted and reported by someone other than the patient. Behind every case report, there is a patient narrative that is typically richer in detail and more personal than what is reported. If one is interested in causal complexity, individual variation and medical uniqueness, the patient narrative should be a better source of information than the case report, if the narrative contains more biographical, personal and contextual information. Rebecca Chandler, medical doctor at the Uppsala Monitoring Centre, writes:

> Professionals working in drug safety also need patient stories. At its very essence, an individual case safety report is a patient story of an adverse experience after using a medicine. Often in pharmacovigilance we focus on numbers and statistics. We discuss Information Component values, proportional reporting ratios, completeness and VigiRank scores. Signal detection using spontaneously reported adverse event data is a hypothesis-generating exercise, a clinical science which is based upon individual reports of suspicions of causality between a medicinal product and an adverse event.
>
> It is logical therefore that clinical stories contained in adverse event reports, complete with details and context, are integral to the development of hypotheses of drug safety concerns. Certain details within the patient story are integral to the building of hypotheses of causality, such as past medical history, concomitant medications, time to onset of symptoms. Other details, if provided, allow us to understand the impact of the event upon the patient's life, their ability – or inability – to manage the adverse event, and even how the patient was treated within the healthcare system.

Pharmacovigilance is more than the identification of causal associations between drugs and new adverse events. It is about creating a culture of awareness of drug safety, and using patient stories to contribute to an evidence base that can be used by physicians and patients to make wise therapeutic decisions. (Chandler 2017: 23)

4.5.2 Narrative as a Tool for Understanding the Causal Story

Many clinicians who are interested in patient narratives are motivated by a philosophical commitment to phenomenology. Ontologically, phenomenology is a version of holism or wholism, and there are a number of overlapping ideas with dispositionalism. Phenomenology has been emphasised and practiced as a methodology by many of the CauseHealth clinicians, including Stephen Tyreman, Anna Luise Kirkengen (Chap. 15, this book), Brian Broom (Chap. 14, this book) and Karin Mohn Engebretsen (Chap. 11, this book). Depending on the version of phenomenology, the narrative should come solely or primarily from the patient, and it should remain as uninterpreted as possible by the clinician. As a clinical methodology, phenomenology emphasises the subjective experiences of health and illness, and also meaning, interpretation, values, existential questions and embodiment. In the case of chronic medically unexplained conditions, phenomenological approaches will have a broader focus suited for uncovering complexity and uniqueness. One example is presented by general practitioner Anna Luise Kirkengen:

> For many years, Katherine Kaplan had been in specialist care due to a long sequence of diseases deemed as separate, different in origin and, consequently, requesting different types of approach. She had been frequently hospitalised with a variety of serious health problems since her late teens. She had encountered physicians in many medical specialties due to what was diagnosed as different diseases in various organ systems. She had been delayed in her studies due to these frequent periods of sickness and was, when finally reaching her graduation, completely incapacitated by chronic states of bad health which could not be responded to with specific treatments any more. Years of medical investment on specialist level were terminated with a referral to a General Practitioner.
>
> In order to understand this disabling process, an analysis of the prevailing concepts of the human body, of diseases, and of medical causality needs to be performed.
>
> When contrasting the "case" depicted above with a biographical account grounded in the "story", a different picture emerges. Katherine Kaplan, the third child of a highly educated and resourceful couple, had been maltreated by both her parents but mostly by her elder brother from early childhood through adolescence and while she was a student of medicine at a Norwegian university. Her parents, defining their abusive acts as deserved punishment, had never realised that their daughter suffered grave and frequent maltreatment by the hands of their son. The on-going threat, embodied as toxic stress in Katherine, increasingly compromised her health preserving systems to the point of breakdown by the time of her graduation.
>
> Anna Luise Kirkengen, 'What if…', CauseHealth blog (https://causehealthblog.wordpress.com/2017/10/17)

4.5.3 Narrative as a Collaborative Tool in Healthcare

A third type of narrative is the collaborative or co-written story, developed in the dialogue between the patient and the clinician. Here the clinician takes a more active role in bringing out, analysing and emphasising different parts of the patient's narrative. In this process, the narrative might change from the individual perspectives of the patient and of the clinician toward a commonly constructed narrative (Low 2017, see also Low, Chap. 8 and Price, Chap. 7, this book). For example, a patient might not think much of how an experience affected their condition, while the clinician might think it is highly relevant. The opposite might also happen. The clinician might not initially understand why the patient mentions something that seems tangential to the medical issue, but then discover from the conversation that this was crucial. In this case, the patient's own narrative becomes re-written as a result of the clinical interaction. This can be an important therapeutic tool, but it is also a tool for uncovering the causal complexity of how the person became ill and what affects the condition in positive and negative ways. The physician then offers the patient some tools to analyse her own subjective experience.

> So why is the notion of story in medicine so foreign to clinicians? Typically we (the clinicians) want the medical truth rather than the human truth (we need both). Medical truth is largely about science, measurement, labelling (diagnosis), and standard ways of treating. These perspectives give us enormous benefits. But in our narrow desire to essentialise, master, know medically, deploy, and instrumentalise, we frequently fail to get close to the patient's unique and individual story, and thus we lose a crucial dimension of the person that may be helping make them sick and keep them sick. If we make it safe for patients to tell their stories, many will do so. We need both the medical perspectives and the stories.
>
> Brian Broom, from the mindbody website: https://wholeperson.healthcare

4.6 To Sum Up…

This chapter offered a philosophical diagnosis of the challenges that medicine is facing, regarding medically unexplained symptoms and complex illnesses. We proposed that a crucial problem comes from applying a Humean regularity theory of causality, in which a cause is understood as something that always provokes the same effect *under ideal conditions*, to the clinical reality, where no ideal condition, or average patient, can ever be found. A dispositionalist understanding of causality proposes instead to start from the particular and unique situation of the single case in order to understand causality. The medical evidence, including causally relevant evidence, must then be generated from the single patient. This includes not only the patient's medical data, but also the patient's condition, narrative and perspective. This is fundamental in order to generate causal hypotheses about the complex situation and all the dispositions that influence the medical condition. Ultimately, evidence from the clinical encounter could assist the design of experiments both in the lab and in the clinics. When possible, one should also use insights from statistical population studies to make decisions about single patients. The best approach to causality, we argue, is to use a plurality of methodologies. We have also explained how, when starting from a dispositional theory of causality, heterogeneity, unexpected results and outlier cases actually represent an epistemological advantage, instead of an obstacle, for the causal enquiry.

References

Anjum RL, Mumford S (2018) Causation in science and the methods of scientific discovery. Oxford University Press, Oxford
Anjum RL, Copeland S, Rocca E (2018) Medical scientists and philosophers worldwide appeal to *EBM* to expand the notion of 'evidence'. BMJ EBM 25:6–8
Berkwits M, Aronowitz R (1995) Different questions beg different methods. J Gen Intern Med 10:409–410
Castaneda D, Popov VB, Wander P et al (2019) Risk of suicide and self-harm is increased after bariatric surgery—a systematic review and meta-analysis. Obes Surg 29:322–333
Chandler R (2017) The patient behind the statistics. Uppsala Rep 77:23
Chew-Graham CA, Heyland S, Kingstone T et al (2017) Medically unexplained symptoms: continuing challenges for primary care. Br J Gen Pract 67:106–107
Copeland S (2017) Unexpected findings and promoting monocausal claims, a cautionary tale. J Eval Clin Pract 23:1055–1061
Craig P, Dieppe P, Macintyre S et al (2008) Developing and evaluating complex interventions: the new Medical Research Council guidance. BMJ. https://doi.org/10.1136/bmj.a1655
de Boer WA, Tytgat GNJ (2000) Treatment of Helicobacter pylori infection. BMJ 320:31
Engebretsen KM (2018) Suffering without a medical diagnosis. A critical view on the biomedical attitudes towards persons suffering from burnout and the implications for medical care. J Eval Clin Pract 24:1150–1157
Engebretsen KM, Bjorbækmo WS (2019) Naked in the eyes of the public: a phenomenological study of the lived experience of suffering from burnout while waiting for recognition to be ill. J Eval Clin Pract. https://doi.org/10.1111/jep.13244
Eriksen TE, Kirkengen AL, Vetlesen AJ (2013a) The medically unexplained revisited. Med Health Care Philos 16:587–600
Eriksen TE, Kerry R, Lie SAN et al (2013b) At the borders of medical reasoning – aetiological and ontological challenges of medically unexplained symptoms. Philos Ethics Humanit Med 8:1747–1753
Go MF (2002) Natural history and epidemiology of Helicobacter pylori infection. Aliment Pharmacol Ther 16:3–15
Greenhalgh T, Hurwitz B (eds) (1998) Narrative based medicine. Dialogue and discourse in clinical practice. BMJ Books, London
Hellhammer DH, Hellhammer J (2008) Stress: the brain-body connection. Karger, Basel
Hume D (1739) In: Selby-Bigge LA (ed) A treatise of human nature. Clarendon Press, Oxford, 1888
Jeffrey JE, Foster NE (2012) A qualitative investigation of physical therapists' experiences and feelings of managing patients with nonspecific low back pain. Phys Ther 92:266–278
Karmali S, Brar B, Shi X et al (2013) Weight recidivism post-bariatric surgery: a systematic review. Obes Surg 23:1922

Lagerros YT, Brandt L, Hedberg J et al (2017) Suicide, self-harm, and depression after gastric bypass surgery: a nationwide cohort study. Ann Surg 265:235–243

Low M (2017) A novel clinical framework: the use of dispositions in clinical practice. A person centred approach. J Eval Clin Pract 23:1062–1070

Mumford S, Anjum RL (2011) Getting causes from powers. Oxford University press, Oxford

Neovius M, Bruze G, Jacobson P et al (2018) Risk of suicide and non-fatal self-harm after bariatric surgery: results from two matched cohort studies. Lancet Diabetes Endocrinol 6:197–207

Rocca E (2017) Bridging the boundaries between scientists and clinicians. Mechanistic hypotheses and patient stories in risk assessment of drugs. J Eval Clin Pract 23:114–120

Rocca E, Anjum RL (2020a) Causal evidence and dispositions in medicine and public health. Int J Environ Res Public Health 17:1813. https://doi.org/10.3390/ijerph17061813

Rocca E, Anjum RL (2020b) Erice call for change: Utilising patient experiences to enhance the quality and safety of healthcare. Drug Saf. https://doi.org/10.1007/s40264-020-00919-2

Solomon M (2015) Making medical knowledge. Oxford University Press, Oxford

Vogt H, Ulvestad E, Eriksen TE et al (2014) Getting personal: can systems medicine integrate scientific and humanistic conceptions of the patient? J Eval Clin Pract 20:942–952

Open Access   This chapter is licensed under the terms of the Creative Commons Attribution 4.0 International License (http://creativecommons.org/licenses/by/4.0/), which permits use, sharing, adaptation, distribution and reproduction in any medium or format, as long as you give appropriate credit to the original author(s) and the source, provide a link to the Creative Commons license and indicate if changes were made.

The images or other third party material in this chapter are included in the chapter's Creative Commons license, unless indicated otherwise in a credit line to the material. If material is not included in the chapter's Creative Commons license and your intended use is not permitted by statutory regulation or exceeds the permitted use, you will need to obtain permission directly from the copyright holder.

Chapter 5
Complexity, Reductionism and the Biomedical Model

Elena Rocca and Rani Lill Anjum

5.1 The Biomedical Model of Illness

Up until the nineteenth century, illness, health and recovery were mysterious matters. Infections, cancer and disease in general were understood as some sort of invading curses, leaving little space for rational treatment. It was only with advances in biological knowledge, such as the development of cell theory, the germ theory of disease and bacteriology, that definite explanations of illness, suffering and death could be formulated. The work of influential scientists and physicians, such as Rudolf Virchow, who is sometimes considered the father of modern pathophysiology, had a revolutionary impact on medical thinking. Virchow introduced the idea that every pathology arises from a damaged cell, which paved the way for the work of Robert Koch and Louis Pasteur, as well as for the development of the first theories of the onset of cancer from a malfunction of host cells. By identifying the origin of disease with a malfunction at the simplest structural and functional level of organisms, the cell, this new paradigm allowed us to conceive of new ways to target the causes of disease, for instance by pharmaceutical interventions.

There is little doubt that the discovery of antibiotics and other drugs changed the course of human history. The consequences of these revolutionary developments, however, go beyond the practical outcomes. There was also a deep change at the cultural and conceptual levels, namely in the way illness and health were understood. The *biomedical model of illness* became the dominant paradigm until the end of the twentieth century.

E. Rocca (✉) · R. L. Anjum
NMBU Centre for Applied Philosophy of Science, Norwegian University of Life Sciences, Ås, Norway
e-mail: elena.rocca@nmbu.no; rani.anjum@nmbu.no

© The Author(s) 2020
R. L. Anjum et al. (eds.), *Rethinking Causality, Complexity and Evidence for the Unique Patient*, https://doi.org/10.1007/978-3-030-41239-5_5

5.1.1 Reductionism in Medicine and Science

Within the biomedical model, an illness is always explained with one or more physical malfunctions at a lower level of organisation. For instance, an infection is explained with the invasion of parasites, a metabolic disorder with a genetic mutation, a psychiatric disorder with an imbalance of neurotransmitters, a speech impairment or a physical disability with neuronal damage and so on. In philosophy, the idea that a complex phenomenon is best understood by analysing its physical parts in isolation, is called *reductionism* or even *physicalism*.

There are many versions of reductionism (for an overview, see Dupré 1993, part II). The version that we are concerned with here is ontological reductionism. According to this view, the world is thought to only have one causally potent level, namely the physical one. Any non-physical phenomenon would then be nothing but the effect of lower-level causes. This idea of reductionism can be seen in the standard hierarchy of the sciences, where physics is the fundament. Above physics is chemistry, then bio-chemistry, biology, psychology and finally on top, social sciences.

Typical for this hierarchy is that for each higher level, more complexity is introduced. Societies consist of people with individual minds and brains, consisting of cells and tissues, which consist of genes, which again consist of molecules and atoms. Ontological reductionism is thus depicting the different levels of existence in a part-whole relationship, where all higher-level phenomena are composed of the levels below (Figs. 5.1 and 5.2).

Medicine might then be placed between biology and psychology, where the complexity of the human mind is treated separately from the body. This distinction is manifested in the way healthcare services are divided into treatment of the psyche (psychotherapy) and treatment of the soma (medicine). A dualist would say that psyche and soma are separate, but equally real, while a reductionist would try to explain psyche as a result of somatic causal processes, such as biochemical

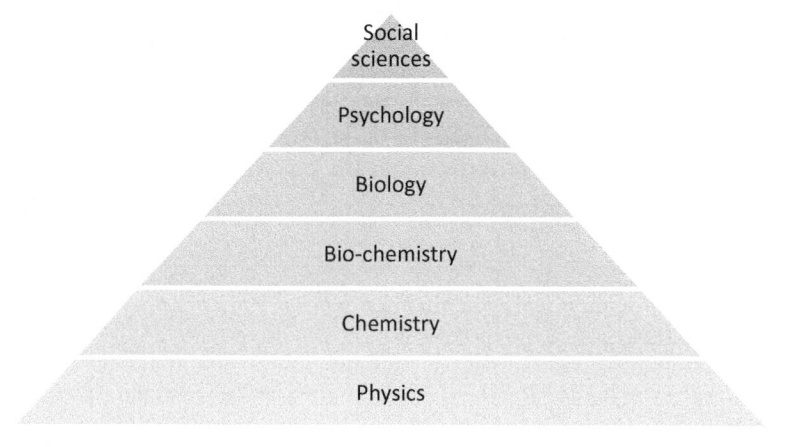

Fig. 5.1 The hierarchy of science

Fig. 5.2 A Venn diagram illustrating a reductionist ontology of wholes and parts

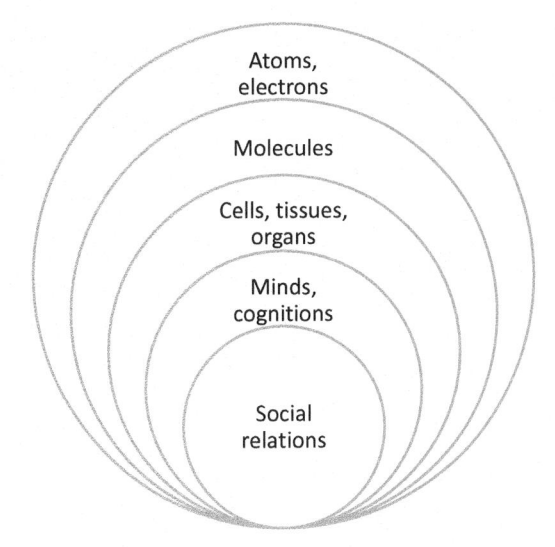

- Atoms, electrons
- Molecules
- Cells, tissues, organs
- Minds, cognitions
- Social relations

interactions. Psychopharmaceuticals, for instance, target psychiatric phenomena as biochemical processes gone wrong. But this is only one way in which reductionism is manifested in the field of medicine.

Another case of reductionism is medicalisation, which is the tendency to treat existential issues and life events as medical ones, and then exclusively as biomedical problems. Instead of feeling shy, one might have social anxiety, for instance, and instead of grieving the loss of a family member, one might be depressed. In some extreme cases, healthy human responses to inhumane life conditions, such as severe childhood trauma or sexual abuse, are defined as psychiatric disorders and approached primarily from a biological perspective (Kirkengen and Thornquist 2013). Once a problem is characterised as a biomedical problem, one will naturally start to search for a biomedical solution to it. As we shall see, this is one of the reasons why the biomedical model has been criticised for being reductionist, and for ignoring important aspects of what it is to be a healthy human being (Getz et al. 2011).

> Simply put...
>
> Reductionism is the philosophical idea that all higher-level (e.g. social, mental or medical) phenomena and processes can in principle be explained at a lower level (e.g. biology, chemistry, physics). Ontological reductionism states that all processes and events must ultimately be the result of physical causes.

The biomedical model brings about some specific ways to understand health, illness and disease. First, illness is always reducible to a physical, biological disease. It concerns purely the physical body, which is seen as analysable into separate parts. This is a mechanistic view of biology, in which parts are not changed by the context, and therefore can be studied in isolation, as one would do with a car engine, for instance. The mechanistic view, together with the dualism that separates body from mind, are deeply set in the Western culture, mostly because of the influential work of René Descartes (1641).

Another idea introduced with the biomedical model is a specific understanding of 'health', which is seen merely as the absence of physical signs of disease. As a result, curing a disease is exclusively a task for medical professionals and medical technology, while the patient is only a receiver of such cures. The biomedical model has an intuitive appeal for many types of conditions, such as bacterial infections. But for a long time this view was taken as generally valid for medical science. Also, the biomedical approach acquired a normative connotation. This means that the orthodox medical thinking has been to consider the most 'scientific' medical intervention to be the one based on the biomedical model. This frame of mind, as we will see, has been widely criticised in the last few decades. Nevertheless, the biomedical model is so influential and deep-seated that it has survived and is still the prevailing view in medicine.

For example, although Attention Deficit Hyperactivity Disorder (ADHD) is diagnosed in children as an alteration of their behaviour, mainstream medicine explains the condition as a neurobiological disorder (cf. DeVreese et al. 2010). As a consequence, the main therapy for ADHD is pharmacological psycho-stimulation, although the precise biological mechanism of such intervention is at the moment unclear. A public health study revealed that children born late in the year were more likely to receive an ADHD diagnosis than those born earlier in the year (Karlstad et al. 2017), suggesting that lack of school maturity in the younger children might be an important causal factor. If so, the symptoms interpreted as an intrinsic neurobiological disorder, ADHD, might actually be the effect of social and contextual factors.

5.1.2 Critical Reflections Concerning the Biomedical Model

The limitation of the biomedical model has been highlighted by many, already since the 1950s. The main criticism is that illness is a condition of the whole person, and treating the patient's bodily parts in separation might alleviate some symptoms without solving the source of the problem. This became increasingly evident with the epidemics of chronic illnesses and metabolic disorders due to an unhealthy lifestyle in Western society, such as diabetes, obesity and cardiovascular disease. Medicalisation without a thorough intervention at the social and psychological level has not been successful in solving these conditions (see for instance Hagen, Chap. 10, this book).

Another criticism of the biomedical model is that it objectifies the patient and reduces them to a passive target of therapy, rather than seen as an active (and the most crucial) actor in healing. As Stephen Tyreman noted, 'patient' is the opposite of 'agent', and a loss of agency is experienced as illness (Tyreman 2017: 277). This suggestion of a lack of agency is why many person centred practitioners insist on avoiding the term 'patient' and prefer 'person' instead. This also signals a holist view where the person is a subject and not primarily seen as an object of illness (the damaged knee, the hurting back, the malignant tumour), as they are within a reductionist view. Finally, the biomedical model fails to account for all the conditions under which a patient is in fact ill, but without presenting any physical or biological sign of dysfunction. Such is the case for the medically unexplained symptoms or syndromes that we discussed in Chap. 4. These are also referred to as medically unexplained *physical* symptoms (MUPS).

Modern medicine, therefore, is faced with a contradiction by which scientific advances and medical technology offer the best opportunities ever, but at the same time an increasing number of patients are over-medicalised, over-diagnosed, become chronically ill, do not find a place in the health system, or feel that they are not met as whole persons in the healthcare system. The biomedical model seems to have played a central role in this development.

In the CauseHealth project (described in Chap. 1), we have seen that people who experience medically unexplained symptoms often become victims of the biomedical model. Since burnout was not generally accepted as a medical diagnosis until the WHO declared it a diagnosis in 2019, individuals with burnout face the societal stigma of being thought of as not 'really ill, but just lazy' (Engebretsen and Bjorbækmo 2019). However, the lack of a biomedical cause also results in a financial burden for patients without a diagnosis, since one might then not qualify for economic compensation in case of long-term sick-leave (Engebretsen 2018). One obvious solution for helping this patient group is to find a symptom or diagnosis that is already recognised by the healthcare system. While this might solve some problems, it might also create some new ones. Psychotherapist and researcher on burnout, Karin Mohn Engebretsen, has seen how the biomedical model motivates the choice of treatment of people suffering from burnout.

> A matter of debate is whether burnout should be considered a distinct medical diagnosis or a form of depression. Recent research has suggested that public health policies should focus on and medically treat one of the core symptoms, which is asserted to be depression. The preferred medication is selective serotonin reuptake inhibitors (SSRI). A problem, however, is that these patients, referred to me by their general practitioners, often complain about worsened symptoms that might be a side effect of the medication they are on.

> Although burnout and depression have similar symptoms, my experience is that there may be substantial differences on the underlying psychological process. I experience the patients grieving a loss quite differently from patients being ill due to an overwhelming life situation. Treating the symptom without any idea of the underlying process in this case might provoke serious trouble. For instance, as some research pointed out, SSRI can lower cortisol levels and therefore worsen the symptoms in stress-induced conditions, possibly through interfering with hypothalamic-pituitary-adrenal axis functioning. Therefore if burnout, as we can reasonably suppose, is a stress-induced syndrome, SSRI can hurt much more than they can help.

To increase the knowledge of burnout as a phenomenon, complementary research methods are required. Person centred healthcare and the Biomedical Model represent two central methodological perspectives that constitute the main camps in contemporary medical and social science. They depict two extremely different ways of studying social phenomena and as such, they may complement each other. Instead of limiting the medical model to specific biological factors, I argue that it is necessary to include the entire human being within a contextual setting to be able to understand the underlying mechanisms. So to improve the healthcare system related to medically unexplained symptoms it is due time to open up for a philosophical reflection on what research questions we need to answer and choose the methodology that will provide these answers.

Karin Mohn Engebretsen, 'Are we satisfied with treating the mere symptoms of medically unexplained syndromes?', CauseHealth blog (https://causehealthblog.wordpress.com/2017/03/27/are-we-satisfied-with-treating-the-mere-symptoms-of-medically-unexplained-syndromes/)

By translating burnout into depression (seen as a biochemical imbalance in the brain), one effectively reduces a complex psychosocial phenomenon to one of its medically accepted symptoms. The serious problem arises when burnout is then treated medically, as if it was actually *caused* by depression.

5.2 The Bio-psychosocial Model of Illness

In the second half of last century, George Engel (1977) proposed a new model to understand health and illness: the bio-psychosocial model. Engel thought that a whole new way of thinking about human conditions was needed. In particular, he found it necessary to acknowledge that not all illnesses are detectable by biological measurements. The bio-psychosocial model embraces all the scientific advances underlying modern medicine, while also highlighting that many conditions cannot be explained by detecting changes at the cellular or molecular level.

For instance, infants who do not receive care and attention from adults might not develop correctly, although all the other physical needs are met. Similarly, patterns of recovery after heart surgery in children are dependent on relationship and communication with family, and even vary depending on whether or not patients have animal companions (Ellis 2012). In these cases, although there are changes at the cellular and molecular level, such changes do not provide a causal explanation for the developmental disruption. Instead, the changes at the lower level of biological organisation are caused by higher level phenomena: in these cases, by social interaction. It is by intervening at this higher level that one can really influence the course of development and recovery.

The bio-psychosocial model aims to introduce to medical and healthcare practice the concept of downward causality, or top-down causality. This is the concept by which causality travels from the higher to the lower level of organisation. In other words, the whole can sometimes cause a change in its parts. Consequently, it is not possible to understand the causal story by analysing the parts in isolation.

> Simply put…
>
> Bottom up causality means that the direction of causality goes from causes at a lower level or organisation to effects at a relatively higher level of organisation, while top down causality goes from causes at a higher level of organisation to effects at a lower level of organisation.
>
> Example: a headache can be caused by hormonal fluctuations (bottom up) or by financial worries in times of economic recession (top down).

There are frequent examples of this in medicine: cases in which the social and psychological well-being of the patient influence her physical state are well known to every clinician. Resilience and motivation of the patient, for instance, are often seen as important ingredients for a medical intervention to yield the desired result. This is why, according to Engel and many others in the last decades, a biomedical model, based on the exclusive treatment of physical constituent parts, cannot provide the correct concept of human suffering and healing, and consequently cannot guarantee effective healthcare. Rather, medicine needs to be informed by a more complete understanding of health and illness, which better depicts the reality of human conditions (Loughlin et al. 2018).

Since its original formulation, the bio-psychosocial model had a considerable influence in medical practice, research and education (Farre and Rapley 2017). However, it has also been the object of controversies and criticism. One problem with this model is that it is vague in the formulation of a method for collecting the relevant biopsychosocial information (ibid.). As an amendment to this, some scholars proposed the phenomenological model, the necessity of understanding the patient and the use of narratives as a clinical tool (Greenhalgh and Hurwitz 1998). We already discussed the value of these tools for the causal enquiry in Chap. 4.

Another criticism is that the bio-psychosocial model is very difficult to put into practice in the current medical community. In the context of today's extremely specialised medical education, which practitioner is trained to catch all three levels in depth? The most realistic picture is the one in which different professionals care about different levels: the physician about the biological, the psychologist about the psychological, and the social caretaker about the social. We see, then, that rather than achieving a truly integrated analysis like the one originally proposed by Engel, the whole person is again separated into different levels of complexity, and such levels are likely to be analysed and treated in isolation from each other (Kirkengen 2018).

5.2.1 Bottom Up and Top Down Causality in Medical Research: Two Views on Cancer Aetiology

We already mentioned how there can be top down or bottom up explanations of causality. The choice of direction will necessarily affect how we think of causality in medicine. Bottom-up explanations typically look for medical causes at the physical level, while top-down explanations will emphasise higher level causes of illness, which might include contextual, psychosocial or ecological dispositions.

Let us look at a concrete example of these competing views from the field of cancer research. There are two competing theories about the onset of cancer. The first, the somatic mutation theory (SMT), is based on bottom up causality, in so far as it identifies molecular entities as the initiating causes of cancer. An early version of this theory was first stated in Weineberg's 1998 book *One Renegade Cell*, where the author proposes that cancer is originated by a single genetically mutated cell. This extreme form of genetic reductionism was reformulated when knowledge about the complexity of cancer was still developing. Today, the SMT theory postulates that there are several types of mutations that can result in cancer, and such mutations are always altering the communication between one cell and its environment. For instance, a carcinogenic cell loses the ability to react to anti-growth signals from the environment. Still, this view is based on bottom up causality, because it identifies some causes at molecular level which modify higher organisation levels, such as the tissue.

A competing view is the tissue organisation field theory (TOFT), formulated in 1999 by Sonnenschein and Soto in the book *Society of Cells*. Causality here works in a top down process: from the tissue to the cell. According to the theory, it is the disrupted tissue that provokes a change of environment and consequently a change in the cell phenotype, from regular to carcinogenic. Cancer, then, would be a developmental illness, because it would derive from a failure in tissue development. Cells would change their phenotype because of a change in their environment (the tissue), and not primarily because of a change in their components (genes, for instance).

Lately, scientific insights on the epigenetic regulation of the cell phenotype are used to understand cancer. For instance, the relapse after cancer pharmacological therapy is not necessarily due to the presence of a randomly mutated cell since before the therapy. This very Darwinian interpretation, where the mutated cell is passively selected, has been questioned (Pisco and Huan 2015). Instead, cancer cells might change their phenotype independently from a genetic mutation, and as epigenetic adaptation to the environment. If this theory is correct, it would highlight that carcinogenic cells adapt actively to changes in the environment.

These different views have profound consequences for the way cancer is understood, but also for how it should be treated. A bottom up view of causality is thus one of the ontological assumptions that tacitly influence medicine both theoretically and practically.

5.3 The CauseHealth Approach: Change Must Start from Ontology

Although the bio-psychosocial model is a step in the right direction, we said, healthcare needs to move beyond it. The model starts with the quest for a more holistic view of health and illness that does not reduce human conditions to merely the sum of their constituent parts. Still it ends up with the fragmentation of the biological, psychological and social dimensions. Is it possible to reach a genuine integration among these three dimensions of humanity, and if so, how?

In the philosophical framework so far presented (see Chaps. 1, 2, 3, and 4), we promote a view by which any genuine change in the practice and the norms of science must start from the revision of the underlying concepts motivating such practice and norms. The way we think about complexity, causality and probability, for instance, is going to affect what we consider to be the best method to study them. Scientists and practitioners often lament the shortcomings of a certain methodology and try to improve it. However, if this improvement does not start from an update of the most fundamental basic assumptions about the reality to be investigated, the methodological improvements will not be very radical.

In other words, a genuine change in scientific and clinical approaches must start from the most fundamental level to the practical one (Fig. 5.3).

Norms of science are here understood as the norms for the 'correct, systematic acquisition of empirical knowledge' (Anjum and Mumford 2018). Note that the norms we refer to in this context are restricted to epistemology: norms about how researchers and practitioners should best collect and process empirical knowledge. These norms differ from the ethical norms, such as for instance autonomy, justice and equality. The reason why we highlight the relevance of epistemological norms of science in this context, is that any practice that falls outside of an established norm of science is likely to be considered unscientific and met with scepticism. We see then that, in order to change practice, one must first make a more radical change

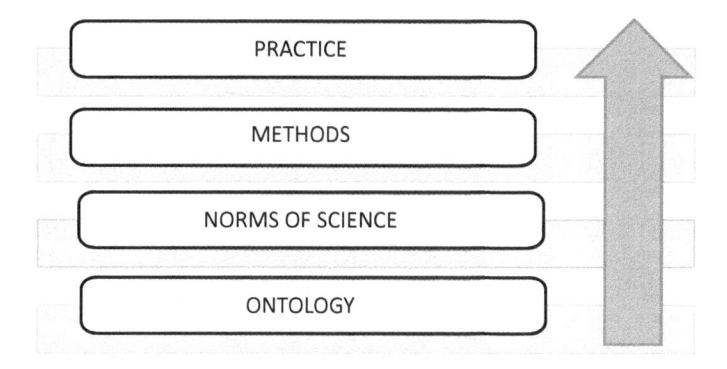

Fig. 5.3 A change in methods and practice must start from a change in ontology

in the norms of science that motivates this practice. Norms of science, in turn, depend on basic ontological assumptions about concepts such as causality, complexity and probability.

Ontology concerns basic implicit assumptions and such assumptions are always present in any type of science, including in medical practice. In the paper 'Philosophical bias is the one bias that science cannot avoid' (Andersen et al. 2019), we argue that such basic implicit assumptions are a necessary prerequisite for any practice informed by science, and for science itself. However, in times of what Thomas Kuhn (1962) calls 'normal science', there is little talk about such assumptions, which remain tacitly and commonly accepted. It is only in times of paradigmatic changes that scientists and practitioners start to talk about ontological basic assumptions and critically discuss them.

One example of such discussion at the ontological level is the criticism of the reductionist and dualistic view of human biology, and the call for a renewal in which human conditions are conceptualised as *complex* and *emergent* phenomena. But even if such critical discussion has been ongoing for many decades, we think that there is still conceptual and fundamental work to do to inform a genuine change in clinical practice. A foundational change in ontology ought to lead to a change in norms and practices, and it should also challenge the way medicine and healthcare is organised, managed and financed. If no such change is seen, it might be because our concepts only *sound* new, but their meaning is defined within the old ontology. For instance, there is general agreement about the complexity and multi-causality of pathogenesis of most illnesses, but there is no general agreement about what it means to say that causality is 'complex'. We will now show how a discussion of this concept can be useful for healthcare and for the clinical encounter.

5.4 What Is Causal Complexity and How Should It Be Investigated?

We have already discussed causal complexity (see Anjum, Chap. 2 and Anjum and Rocca, Chap. 4, this book). But we have not said much about what we mean by such complexity. Does it simply mean that there are multiple causes? This seems to be what the term 'multifactorial' indicates. We will now present two philosophical views on complexity. Depending on which of these views one assumes, different norms, methods and practices will follow.

A common way to think about a complex whole is to see it as the sum of many parts, connected by intertwined causal interactions. For instance, the human genome can be seen as a complex whole in the sense that it is constituted by a large number of functional units, the genes, which are linked by an intricate net of causal interactions. One gene can cause or prevent the expression of many other genes, and can in turn be regulated by a number of different others. In order to understand the causal role of single genes within the genome, scientists then isolate the gene from the genome one by one and study their sequence and their function in different contexts. This practice of isolating one causal factor from its normal complexity is a dominant epistemological norm in science when studying causality. What does this tell us? First of all, it tells us something about how we think of causality, as something that is best established by looking at the behaviour of a single factor in isolation from contextual interferences. Secondly, it reveals something about how we understand complexity.

5.4.1 *Mereological Composition*

We can illustrate this first view of a 'complex whole' with a simple example. Imagine a construction made with Lego bricks. Depending on the shape of the brick, each brick can bind to one or more of the other bricks. Together the bricks can combine to form different wholes, such as a castle or a ship. But crucially the individual bricks do no change from taking place in the different constructions. The bricks maintain their original properties throughout. We will call this view of complexity, in which a whole does not induce a change in its parts, *mereological composition*.

Mereological composition is an ontological thesis about how parts relate to wholes and to the other parts within that whole. Crucially, the parts are thought to maintain their properties and identity when combined with other parts to form the whole. Mereological composition might also entail the view that wholes can be decomposed into their parts. This is for instance how a car works. One can put the parts together to make the car and then one can take it apart. The parts of the car will be the same before the composition and after the decomposition. How does this relate to healthcare and the clinical encounter?

> Simply put…
>
> Mereological composition here means that the whole is the sum of its parts and that, throughout the process of composition and decomposition, the parts remain unchanged within the whole.
>
> Example: a car engine is produced by the mereological composition of its parts.

A criticism of the biomedical model has been that it sees a person in the same way – as a whole that is best understood by studying and treating its individual parts in separation: the liver, the heart, the lungs, and so on. The bio-psychosocial model might end up with a similar assumption if we have to study the biological, psychological and social causes separately and then add up the results. This is not primarily a shortcoming of the bio-psychosocial model, but of the scientific methodology of isolating and separating each causal factor and studying them independently of their natural context. This scientific approach comes from the ontological assumption of mereological composition.

A genuine whole-ist should not accept a compositional view of complexity, even if the complexity consists in biological, psychological *and* social parts. What does this mean for the clinical encounter? Is the orthodox scientific approach, based on a mereological idea of complexity, the only feasible option? What is the role of biomedical knowledge in the wholistic clinical encounter? Immunologist and psychotherapist Brian Broom has explored this question in depth during his career. He writes:

> So, does this mean that the whole person-oriented biomedical clinician should, in addition to the normative clinical requirements of his discipline, somehow become a skilled psychotherapist, psychologist, social worker, spiritual advisor or whatever, and be required to perform elaborate, expert, systematised assessments normative within each of those disciplines? This is nonsensical and impossible. Nobody can attend to all of this.
>
> The usual solution is of course the multidisciplinary team. But many of these are pass-the-parcel scenarios where each discipline functions narrowly according to the pattern ordained by the modern biomedical model as expressed in each individual discipline. In sum it usually manifests as an additive framework of highly expert clinicians, patients seen from multiple narrow perspectives, a dualistic concept of disease, and a lack of attention to the highly nuanced individual personal life experiences and subjectivity factors or stories that contribute to the development and perpetuation of disease.
>
> Apart from that critique, I actually value multidisciplinary teams, but believe that each of the practitioners in the team need to be functioning in a whole person way. This is possible whatever one's discipline. By adopting a whole person approach each clinician can do a great deal to enhance healing without feeling overwhelmed.
>
> Brian Broom, 'Imagination and its Companions', CauseHealth blog (https://causehealth-blog.wordpress.com/2017/07/03)

According to Broom, multi-disciplinary teams of clinicians, where each clinician adopts a 'whole person approach', is the way forward to enhance healing. The question then becomes: how should clinicians successfully and genuinely embrace the whole person approach? This cannot be done without revising perhaps the most foundational premise of medical research and practice: the way we understand complexity.

5.4.2 Genuine Complexity and Emergence

There is another way to understand complexity than as mereological composition. This is what we will call 'genuine complexity'. On this view, complex wholes consist in parts that interact with each other in a way that also influences and alters the parts themselves in the process. As parts of a whole, the parts are no longer clearly separated in a way that they can easily decompose and compose into new wholes, with their identity intact. Instead, the interaction of the parts within that whole is what will give the identity to each part. Outside the context of the whole, the parts would not be that particular part with those particular causal powers or dispositional properties. Their causal role is given by their place and interaction *as part of that particular whole*. The molecule of DNA, for instance, has a specific causal power because it is part of a whole cell, and of a whole organism. DNA extracted from the cell has no causal power, and degrades in a short time.

Recall the concept of a mutual manifestation partner (see Anjum, Chap. 2, this book): the same causal disposition in a different context, or whole, manifests differently because it interacts with different manifestation partners. If this is the case, as dispositionalism assumes, then a complex whole can never be completely understood by observing its parts in isolation. On this view, the *interaction* among the constituent parts and the whole is as important for the causal inquiry as the parts themselves. This is also because the result of such interactions cannot be seen as mere composition. Instead, the whole is the result of a continuous and complex process where the parts that interact lose their prior identity along the way. The whole is therefore more than the sum of its parts, or even something else entirely. We can say that the whole is an *emergent* phenomenon. From the perspective of dispositionalism, ontological emergence is the view that the whole has new properties and new causal powers as a result of the causal interactions of its parts, where the change also happens in the parts during this process (for more details on dispositionalist emergence, see Anjum and Mumford 2017). A simple example of this could be water, which has a number of causal powers that are not found in its atomic components. Water is thus the result of a process of change that happens when the atoms interact to form the molecule.

> Simply put…
>
> Emergence happens when there are new properties and causal powers of wholes in virtue of causal interactions among their parts. The whole is then more, or something else, than the sum of its parts.
>
> Example: sodium chloride is composed by sodium and chlorine, yet its properties are completely different from the properties of its components.

This way of thinking about complexity is more common in the discipline of ecology, where the interaction between a species (the part) and an ecosystem (the whole) changes both. A beaver, for instance, modifies its surroundings by building a dam. But at the same time, the surroundings modify the beaver by natural selection. For this reason, ecology studies the interactions between species and ecosystems, and would not be interested in studying a species in captivity, isolated from its natural context.

Can the ecological perspective add something to healthcare and to the clinical encounter? We think so. Up until now the biomedical model has been dominant in medicine, but with a limited understanding of biology, taken from molecular biology, biochemistry and physiology. Although this knowledge is necessary for clinical work, we urge that it not sufficient for it. Dispositionalism suggests that medicine would benefit from an ecological turn. Such a turn toward an ecological perspective in medicine would place much more emphasis on understanding human biology as genuinely interactive, and on investigating how biological processes are integrated with human context and lived experience. In the words of osteopath Stephen Tyreman:

> Understanding what person-centred means is much more complex and multi-factorial than I once assumed. It is not merely a question of considering a person's individual needs and concerns and putting them first. It is recognising that human beings face up to the challenge of illness, pain and disability differently from how we might understand and seek to correct a fault in a car, say. (Tyreman 2018: 2)

5.4.3 Practice Is Motivated by Ontological Bias

We have presented two views on complexity: mereological composition and genuine complexity or emergence. Mereological composition was illustrated by the Lego bricks, where the parts combine to compose different wholes, but without any change to the parts themselves. Genuine complexity was the holist alternative, where the whole is an emergent existence in which the parts interact and change each other. The whole then has properties that are different from the composition of the properties of its parts. How does our ontological assumption about complexity affect scientific and medical practice? We can show this by applying the two philosophical perspectives to human pathogenesis.

Consider an autoimmune disease that might have biological causes (e.g. genetic predisposition) as well as psychosocial causes (e.g. lifestyle or emotional stress). Under the assumption that complexity is compositional mereology, the intertwining of different types of causes would represent a challenge to understanding causality. This is why fragmentation of different causal contributions, and their evaluation in isolation, is a well-established norm for scientific inquiry into causality. Genetic predisposition, for instance, might be tested by genotyping of patient groups, or by looking at the susceptibility of a lab animal strain with the genetic mutation(s) we want to test. The causal role of emotional stress might then be investigated through case studies, cohort studies, or other types of clinical studies. In the end, the results from different studies can be added together to give us the causally complex result.

This way of thinking about the scientific approach in medical research also influences medical practice, when dealing with complex medical conditions. Current medical practice aims to combine biological causes with psychosocial causes of illness. But also here complexity seems to be understood and handled according to the biomedical orthodoxy.

ONTOLOGICAL BIAS
Complexity is mereological composition of changeless parts.
⇓
NORM
The investigation into causal complexity starts with separation and isolation of each component.
⇓
PRACTICE
Fragmentation of causes into separate biological, psychological and social spheres is the starting-point for medical inquiry into complex conditions.

The introduction of person centered healthcare was partly motivated by the anti-dualist and anti-reductionist view that health and illness must be understood as belonging to the person, not to one or more bodily part. From this perspective, the mereological composition view is a simplification of the human condition. Such simplification might at times be useful in some contexts, but it can be dangerous, too. In the words of Marie Lindquist, director of the WHO collaborating Uppsala Monitoring Centre for International Drug Monitoring:

> Our ability to quickly categorise things around us is a basic instinct, a survival mechanism, and it was essential in a time when the ability to quickly identify danger was a matter of life or death. By classifying and grouping things, we make a complex reality more manageable. The problem is if we categorise in a way that is confining and excluding, and reduces reality too much – a simplistic reductionist approach easily leads to stereotyping, which can be anything from irritating to seriously damaging. (Lindquist 2018: 2)

Dispositionalism replaces mereological composition with emergence, which we take to be a type of genuine complexity. Genuine complexity is a result, not only of the composition of different parts, but also of their mutual interactions. For instance, the outcome of organ transplantation depends on how well the new organ interacts with the rest of the body. By focusing on such causal interactions, we also need to replace the epistemic norm for how to deal with causal complexity, scientifically and in medical practice.

ONTOLOGICAL BIAS
Complexity is genuine and emergent.
⇓
NORM
The investigation of causal complexity starts with observing the whole, and the interaction of different elements as parts of that whole.
⇓
PRACTICE
Whole person centred practice and patient narratives are the starting-point for medical inquiry into complex conditions.

We see, then, that it makes a difference both to scientific norms and practices how we understand complexity, ontologically. Ontology thus influences the norms and practices that define a scientific discipline. By questioning the philosophical biases of our methods and practices, one can also challenge what counts as scientific practice.

5.5 We Need an Ecological Turn in Medicine and Healthcare

We have argued that, once we acknowledge a phenomenon as genuinely complex, we also need to make some epistemological commitments for how to study that phenomenon scientifically. The most important is the fact that, if we aim to understand (and heal) living beings, we need to start by understanding complex *interactions*. This is not to deny the obvious: we all consist of parts (a liver, a heart, a brain, and so on), and there is much general and indispensable knowledge to gain about organs, tissues and cells by observing their behaviour in experimental and isolated contexts. The tricky part comes when we need to make use of such knowledge for the treatment of the whole person. In that case, it is necessary to keep in mind that although whole persons consist of parts, the identity of such parts is defined by their *interactions*: not only their interactions with the rest of the body, but also with the person's context, her history and her lived experience.

Not many would disagree with this, we think, when asked to reflect upon it. Denying it would be like comparing the human body to a machine, in which single parts interact without ever changing as a result of such interaction. This would entail that human life could be disassembled into its parts, and then re-created by re-assembling them. An absurdity, we might say. Yet, current practice and thinking in the medical profession tend to drift toward the mereological, or mechanical, simplification of human biology. Talking about a 'heart condition', 'irritable bowel' or 'skin disease' has a practical function, but in the long run it does make us think of illness as belonging primarily to a part of the body. As a consequence of this, we might come to think of illness as something that is curable by treating that part alone. In the clinical reality this becomes highly problematic. Here, co- and multi-morbidity are the rule rather than the exception, and most if not all medical conditions are at least partly caused by contextual factors. Patients with complex conditions will then have to be treated by different specialists for different health complaints, and the full picture is often lost.

> While co- and multi-morbidity is the norm in medicine, clinical guidelines are for individual illnesses. The 'guidelines mentality' often results in a situation that has been referred to as *silo medicine*... where each diagnosis has its own expert groups, patient organisations, industry sponsors and clinical guidelines. Diseases are then treated as wholes ("disease holism"), while patients are treated as composed of parts ("patient compositionality"). (Anjum 2016: 423)

What do we need, then, in more practical terms? How can we move toward an ecological turn in medical care: that is, one that understands health and illness by starting from interactions both with the physical and with the psycho-social? One important step is suggested by Getz et al. (2011) in their paper 'The human biology – saturated with experience'. Here they argue that the medical profession must acknowledge lived experience, meaning, and interpretation as not just 'side information' about the patient and her preferences, but as actually and physically influencing human biology.

> Long-term overtaxation of the physiological adaptability of human beings may lead to health impairment. This phenomenon, called «allostatic overload»…, is a consequence of physiological «wear and tear» due to strong and/or persistent threats to an individual's existence or integrity (the word encompasses both mental and physical aspects). The human body's reaction to stressors, which in our culture can be classified as physical (e.g. under-nourishment, overfeeding, malnutrition, pollution, lack of sleep, lack of exercise, infections, noise) prove to converge at the same biological «level» as stressors we would classify as psychosocial (e.g. a life characterised by threats, neglect, abuse, poverty or overwhelming caregiving burdens): Both categories of stress can contribute over time to the development of autonomic dysfunction, changes in the immune system, chronic low-grade inflammation, endocrine disruptions and accelerated cell aging, measured as telomere shortening… (Getz et al. 2011: 684)

Accordingly, if we understand human biology from an ecological perspective, we cannot treat health or illness solely as the results of internal physiological or biochemical processes. Instead we should look for internal as well as external and contextual causes that might influence a person's health in a positive or negative way. For health, all levels of nature are united within one single patient: physiological, biological, psychological and social. All these levels affect and are affected by health and illness. Ecologically, we should also expect that these different influences will interact in nonlinear ways, and that what happens in one context cannot automatically be transferred to another context. Consequently, human biology cannot be understood without including, and even starting from, the higher level of complexity, in a top-down way.

From this ecological perspective, we might ask how efficient it is for medicine to be organised into separate specialisms, each with their own clinical guidelines (see Copeland, Chap. 6, this book for a more detailed discussion of guidelines).

5.5.1 Whole Person Healthcare in Practice

Immunologist and psychotherapist Brian Broom started a multidisciplinary centre in New Zealand in 1987 to offer what he calls whole person healthcare. The idea is to understand health and illness in a way that genuinely overcomes both reductionism and dualism, emphasising the link between life experiences and physical illness. The main goal is that patients, once they feel met as a whole and not as the sum of biological parts, will start to explore their own illness in a whole-istic way, in light of their lived experience. Broom and his team use patient stories, together with the medical perspective, as an essential tool to understand and treat chronic illness. This is how he describes his practice.

> Asking for a story may seem a simple matter but the implications are hugely important. And, in listening to a story, imagination is important… An example: A clinician asks a patient for her 'story', about what happened in her life when her symptoms started many years ago when she was 18 years old. This question is actually asked because the clinician assumes to some degree that the patient is a unitive whole, **and** that the patient's life experience (at 18) may be very relevant, **and** that opening this up may help the person get well in some way. Simply asking the question rests on really serious foundations. For me it entails

the conviction that mind and body co-emerge together from the beginning of life, **and** therefore it is natural to imagine the story being part of the illness. It follows that there may be therapeutic potential in knowing the story. This is the paradigm of whole person care. The paradigm allows the clinician to imagine 'something' important in the story around age 18, and to ask the simple question.

Brian Broom, 'Imagination and its Companions', CauseHealth blog (https://causehealth-blog.wordpress.com/2017/07/03)

Since biological conditions are part of meaningful humans, they are also meaningful. Broom talks about 'meaning-full disease', and his books collect many astonishing examples of the identity between mind and body, which he calls 'mindbody'. The whole person website offers numerous resources, including a tool called 'illness explorer', in a version for patients and one for practitioners. This tool is meant to guide the understanding of the disease as meaningful, and of meaning as embodied (https://wholeperson.healthcare). In Chap. 14 of this book, Brian Broom gives a detailed historical account and description of his practice.

5.6 To Sum Up…

We have seen that the biomedical model of illness relies on a dualistic and reductionist view of the human condition. Although intuitive and attractive for medical research and especially for some types of pathologies, this perspective has nevertheless been criticised for almost a century. The problem is that the biomedical model fails to see illness as a matter of the whole person, in that it overlooks the importance of social, lifestyle and psychological factors in the onset of complex and chronic disorders. Despite decades of criticism of the biomedical model, the current state of the art suggests that it is still predominant in medicine. Proposed alternative views, such as the bio-psychosocial model, still convey a fragmented perspective on human biology, and consequently a fragmented medical care. To face this problem, we propose a re-discussion of the foundational concept of *complexity*. While this concept is widely used, its meaning and interpretation usually remain implicit. Although complexity and emergence have become important words in medicine and clinics, we think that there is not enough reflection on what they really mean.

In this chapter, we have shown that a mereological view of complexity, in which complexity is seen as composition of multiple parts, motivates an investigation that starts from the separation of causal factors, and their investigation in isolation. In contrast, we propose what we call 'genuine complexity', in which the parts of a whole not only compose and interact, but also change each other through such interaction. This, however, requires that we start an investigation from the higher level of complexity: by observing the whole. At such a level, indeed, it is possible to focus on interactions between context, lived experience and physical body parts. Several clinicians, globally, are pushing for a change in this direction. An ecological shift in medicine, we argued, will be not only necessary, but also unavoidable, if we acknowledge that human biology is *genuinely* complex and we truly reflect on the meaning and implications of this.

References and Further Readings

Andersen F, Anjum RL, Roca E (2019) Philosophy of medicine: philosophical BIAS is the one bias that science cannot avoid. elife. https://doi.org/10.7554/eLife.44929
Anjum RL (2016) Evidence-based or person-centered. An ontological debate. Eur J Pers Cent Healthc 4:421–429
Anjum RL, Mumford S (2017) Emergence and demergence. In: Paoletti M, Orilia F (eds) Philosophical and scientific perspectives on downward causation. Routledge, London, pp 92–109
Anjum RL, Mumford S (2018) Causation in science and the methods of scientific discovery. Oxford University Press, Oxford
De Vreese L, Weber W, Van Bouwel J (2010) Explanatory pluralism in the medical sciences: theory and practice. Theor Med Bioeth 31:371–390
Descartes R (1641) Meditations on first philosophy. In: Cottingham J, Stoothhoff R, Murdoch D (eds) The philosophical writings of Descartes, vol 2. Cambridge University Press, Cambridge
Dupré J (1993) The disorder of things. Metaphysical foundations of the disunity of science. Harvard University Press, Cambridge
Ellis GFR (2012) Top-down causation and emergence: some comments on mechanism. Interface Focus 2:126–140
Engebretsen KM (2018) Suffering without a medical diagnosis. A critical view on the biomedical attitudes towards persons suffering from burnout and the implications for medical care. J Eval Clin Pract 24:1150–1157
Engebretsen KM, Bjorbækmo WS (2019) Naked in the eyes of the public: a phenomenological study of the lived experience of suffering from burnout while waiting for recognition to be ill. J Eval Clin Pract. https://doi.org/10.1111/jep.13244
Engel GL (1977) The need for a new medical model: a challenge for biomedicine. Science 196:129–136
Farre A, Rapley T (2017) The new old (and old new) medical model: four decades navigating the biomedical and psychosocial understanding of health and illness. Healthcare. https://doi.org/10.3390/healthcare5040088
Getz L, Kirkengen AL, Ulvestad E (2011) The human biology-saturated with experience. Tidsskr Nor Laegeforen 131:683–687
Greenhalgh T, Hurwitz B (eds) (1998) Narrative based medicine. Dialogue and discourse in clinical practice. BMJ Books, London
Karlstad Ø, Furu K, Stoltenberg C et al (2017) ADHD treatment and diagnosis in relation to children's birth month: Nationwide cohort study from Norway. Scand J Public Health 45:343–349
Kirkengen AL (2018) From wholes to fragments to wholes—what gets lost in translation? J Eval Clin Pract 24:1145–1149
Kirkengen AL, Thornquist E (2013) When diagnosis makes us blind. Tidsskr Nor Legeforen 133:1466–1468
Kuhn T (1962) The structure of scientific revolutions. University of Chicago Press, Chicago
Lindquist M (2018) Director's message. Uppsala Rep 78:2
Loughlin M, Mercuri M, Parvan A et al (2018) Treating real people: science and humanity. J Eval Clin Pract 24:919–929

Pisco AO, Huan S (2015) Non-genetic cancer cell plasticity and therapy-induced stemness in tumour relapse: 'what does not kill me strengthens me'. Br J Cancer 112:1725–1732

Sonnenschein C, Soto AM (1999) The Society of Cells: cancer and control of cell proliferation. Springer Verlag, New York

Tyreman S (2017) How to work with the anthropo-ecological narrative in clinical use. In: Mayer J, Standen C (eds) Textbook of osteopathic medicine. Elsevier, Munich, pp 277–283

Tyreman S (2018) A personal reflection on person-centred care and the role of stories in healthcare. J Osteop Med 28:1–3

Weinberg RA (1998) One renegade cell: how cancer begins. Basic Books, New York

Open Access This chapter is licensed under the terms of the Creative Commons Attribution 4.0 International License (http://creativecommons.org/licenses/by/4.0/), which permits use, sharing, adaptation, distribution and reproduction in any medium or format, as long as you give appropriate credit to the original author(s) and the source, provide a link to the Creative Commons license and indicate if changes were made.

The images or other third party material in this chapter are included in the chapter's Creative Commons license, unless indicated otherwise in a credit line to the material. If material is not included in the chapter's Creative Commons license and your intended use is not permitted by statutory regulation or exceeds the permitted use, you will need to obtain permission directly from the copyright holder.

Chapter 6
The Guidelines Challenge

Samantha Copeland

6.1 The Tension Within

6.1.1 Evidence Based Medicine and the Rise of Guidelines

Part of the reason behind the popularity of evidence-based medicine, or EBM, in the last few decades has been its promise to standardise care. When Guyatt et al. (1992) introduced the concept and practice of EBM in medical education, it was in direct response to the 'old ways' of medicine. As they depict it, medicine was a profession where well-known doctors were looked to as the authorities about healthcare: expertise came from experience, and was demonstrated by the admiration of one's peers. Standards of care could vary widely from hospital to hospital, and new evidence about what practices and interventions worked best was rarely taken up by institutions (Guyatt et al. 1992). The appeal to the 'best evidence' as the driver of the standard of care meant that everyone could have access to the reasons behind why things were done in a certain way; such reasons ought to be objectively good reasons, rather than merely authoritative. Patients could expect the same quality of care from every practitioner, and doctors would know the best thing to do for their patients.

For many reasons, this ideal was hard to realise in the world. First, because individual doctors could not possibly be expected to study all the available evidence for any given intervention, and so assessments of what is the best evidence and the recommended strategy for care had to be centralized. Organizations like Cochrane

---

In this chapter, we look at guidelines as an example of a medical institution and practice that would be affected by a shift in ontology from the current biomedical models toward a dispositional model of causality.

S. Copeland (✉)
Ethics and Philosophy of Technology Section, Delft University of Technology,
Delft, The Netherlands
e-mail: s.m.copeland@tudelft.nl

gained in authority, becoming the sources of 'systematic reviews' that did this work on behalf of practitioners. Second, because standardizing care takes more than everyone having access to the best evidence about which practices are optimal. It takes governance, and the organization of institutions with many levels of care and expertise, with different needs and different jobs. Furthermore, medical institutions, in many nations, are public institutions. Consequently, their governance is the concern of national governments, too, and standards must be set for ethical and political reasons. As a result of these and other influences, healthcare guidelines became a resource for governments, medical institutions, and practitioners alike. They are a means for standardizing care, they are meant to communicate which practices and decisions the best evidence says is optimal, and they are both a guide to uncertain practitioners and the means to inform patients of their options.

6.1.2 Guidelines in Practice

However, given what has been said in this book so far, it seems there must be a conflict between standardizing care, as guidelines seek to do, and using the best available evidence to provide optimal care for the individual patient, as practitioners who use guidelines seek to do. That is, the possibilities offered by dispositional causality for understanding health, disease, and the effectiveness (or not) of medical interventions, point to a tension in the goals of guidelines, under the EBM interpretation of what they can and should do.

This tension is found at a larger scale as well. That is, governments take a public health approach to medicine, and for that purpose, statistics are indeed ideal. Clinicians on the ground in medical care, however, find themselves dealing with individual patients in unique situations. There is not always a box to tick to describe their patient, or what they have done to care for them, on the standard forms they are required to fill out. Insofar as EBM encourages the standardization of care and the quantification of medical evidence, it caters to the public management approach to healthcare that is currently frustrating many healthcare practitioners.

In this chapter, we take a look at this tension, and give some ideas about how we can use the philosophical tools introduced so far to resolve it. Specifically, we focus on how this tension has manifested itself in the debate about healthcare guidelines. Healthcare guidelines in one sense exemplify the standardization of medical care. They present a predetermined list of options, they identify and plan a treatment course for a patient according to the category that fits them best, and they are generated by interdisciplinary committees using the best available evidence. However, they are simultaneously an honest attempt to provide guidance to the practicing, individual clinician, who hasn't the time to delve into all the relevant evidence herself, and must rely on others to assemble the best available evidence into a format she can use in her daily, busy practice. Thus, the guideline can be both a hegemonic force, representing consensus and conformity, and also a helpful aide when a clinician is truly in need of information and guidance—they are meant to be both rule and resource. This dual role creates tension in the clinical encounter because following the rule is not always the best thing to do in a particular situation, and so in many cases what the clinician needs is a resource that is not at all a rule. Guidelines, then, to truly aide the clinician, must be something other than rules.

These tensions can be truly problematic, when clinicians are asked or are otherwise inclined to believe it is better to follow guidelines even when their expert judgment would suggest a different path.

> What singular question could be more pressing for clinicians today: how do we prepare the way for the return of the person in contemporary healthcare amidst rife healthcare commodification and the mechanical one-size-fits-all approach that is EBM? …In clinics and hospitals the world-over, the narrative, the personal and the biographical person is often found bullied and threatened by this hegemony that is contemporary EBM; healthcare is de-personalised.
>
> Chris Worsfold, 'Learn to stop worrying and love evidence based medicine', CauseHealth blog https://causehealthblog.wordpress.com/2016/02/07

6.2 Guidelines and Tramlines

A point that was frequently brought up in discussions during CauseHealth events was the worry that guidelines were often regarded as rules to be followed, rather than as guidance about what care options are available and appropriate. This phenomenon is captured by the frequently used phrase 'Guidelines not tramlines'—rather than a set track, the following of which is not a choice but rather given, guidelines ought to offer choices to clinicians and patients. The tendency to treat guidelines as though they were tramlines is, as suggested above, part and parcel of the public management approach to medicine, and presents clinicians with a conflict of interest: do they risk reprimand and even litigation, if they follow their expert clinical judgment to act in a way not advised by the guideline? Osteopath Stephen Tyreman thought this was a problem in health care, and promoted Values Based Practice (VBP) as a remedy:

> VBP is being given further support by the recent Montgomery judgement [UK] in which 'doing the right thing' for a patient is not just following the guidelines or taking account of risks as a percentage, but taking into account 'what matters' to the patient, what his/her values are. This involves a different kind of decision-making and can't be achieved by reading out all the risks, or ticking a box that says we've told the patient the implications of the procedure being recommended. It involves asking how this treatment might impact on the patient's life and what matters to them. This requires a proper dialogue. What the regulators realise they are up against in supporting this is that practitioners will not be able to follow a safe procedure—an algorithm or guideline—that will guarantee they do the right thing; they must make a judgment weighing up a range of fact and value-based criteria, which of course is what David Sackett originally intended EBM to be (e.g. Sackett et al. 1996). The response from some practitioners is to say, 'we haven't got time to do all this deep and meaningful discussion stuff, just tell me what the right thing to do is.' Or, 'what if I make the wrong decision? I don't want to get into trouble.'
>
> Stephen Tyreman, 'Standards for regulation', CauseHealth blog https://causehealthblog.wordpress.com/2016/12/15/standards-for-regulation/

6.2.1 Guidelines and Evidence Based Policy

One of the issues is the content of guidelines themselves. When designed, for instance, to effect changes in the management of healthcare, guidelines indeed present as rules to be followed. Unless everyone follows such guidelines, the changes won't happen. For example, there may be a guideline about how much time to spend with a patient on average. Or, there may be specific guidelines about what steps to take with patients who have certain diagnoses—which tests ought to be done, which specialists engaged, or even how long a session of cognitive behavioural therapy is allowed. When such guidelines are tied to billing processes in the healthcare system, they can become constraints on what is possible for clinicians to do or to offer their patients, rather than guidance about what might help. One could argue, however, that guidelines that focus on management issues really ought to be treated as tramlines, since they require maximum compliance in order to be effective. As an example, expensive experimental therapies should not be used up on patients who cannot clearly benefit from them, for instance because they are in a terminal phase of life. It is hard to think of a good reason to deviate from this managerial guideline. But in the example above, where the length of a cognitive therapy session is mandated rather than suggested, management-focused guidelines seem to cross over into the territory of the clinical encounter.

Another example of this was given at the Guidelines Challenge Cause Health conference by speaker Hálfdán Pétursson (Pétursson et al. 2009; for a review of talks at that conference, see Anjum et al. 2018). He discussed the problems encountered by general practitioners (GPs) who are trying to follow the guidelines for preventing cardiovascular disease in Norway. If the guidelines are followed as written, it would require more hours of work than the entire GP workforce in the country could put in, just for the one task of preventing cardiovascular disease in those who are deemed at risk. As this example shows, not all aspects of healthcare are manageable through rules, and sometimes implementing the best evidence in an ideal way is simply impractical.

Further, when we bring what has been said so far in this book into play in this discussion of the problem of guidelines, we can see that there are also ontological and epistemological reasons behind the problematic but common view that guidelines present tramline-type rules rather than mere guidance. We explore these reasons in the next two sections, as well as how the philosophical framework of dispositionalism might offer a resolution to these guidelines-related challenges.

6.3 The Ontology of Guidelines

The ontology of guidelines can mean two different things. It can mean the nature of a guideline itself—its function, its form, and what goes into developing it—or it could point to the ontological assumptions that shape guidelines as they are developed and used. It is the position of CauseHealth that these two meanings of ontology are intertwined. How we think guidelines ought to be used reflects, and in turn is reflected in, what we think guidelines are about, and why we think they can be effective.

One might say that healthcare guidelines exemplify a utilitarian approach, for instance. Utilitarians like Jeremy Bentham argue that the best (moral, correct) action is the one that will result in the greatest utility (pleasure, or happiness) for the greatest number of people overall. So, a frequentist utilitarian would argue that guidelines should be designed as rules to follow that, overall and if everyone follows them, are most likely to lead to the best possible outcome for the highest number of people. This works well at a population level: consider, for instance, the importance of 'herd immunity' as a reason for making a vaccination mandatory for all healthy members of a population. But this, as we have seen, is not the same as obtaining the best outcome for a particular person at a particular time, which is generally the goal in the clinical encounter.

> Simply put…
>
> Utilitarianism is a form of consequentialism, the ethical framework that assesses moral action according to its impact on others. That is, we need to consider the consequences of the action, to know if it is the right one, when compared with the consequences of other possible actions. Bentham popularized the idea of 'utility' as a way of measuring and comparing the impact of possible actions—actions with greater utility caused more pleasure and less pain overall.

The nature of guidelines is just what we discussed in the last section: are guidelines rules to be followed, or are they collections of good advice and a presentation of relevant options to a clinician and her patient? In order not to treat guidance for the clinical encounter as a managerial rule, we have said that it is important to avoid seeing guidelines as tramlines. But more than this, it is also important to conceive of the function, form, and creation of a guideline in the right way, so that they are developed and used correctly and effectively. The difficulty is in resolving the tension between the need for flexibility, to allow for the particularities of the clinical encounter to influence decisions about care, and the need for standardization of access for all patients to quality care.

6.3.1 Logically Speaking, Guidelines Cannot Be Rules

Anjum and Mumford have written on the nature of guidelines (2017), arguing that guidelines must, logically speaking, be mere guidance rather than hard rules. Even if a guideline is effective as a rule, it still doesn't help the clinician decide what to do in the individual case. That is, the clinician must, in each case, still weigh up whether the guideline offers the best path or options for treatment for the particular patient in care. Thus, the guideline cannot be designed as a rule that can be universally followed, given a diagnosis or situation; because its relevance can be questioned in reflection upon the particularities of an individual patient, the guideline itself offers a choice, rather than a rule. For instance, in Chap. 4 of this book, Anjum and Rocca describe the case of guidelines for morbid obesity. While current guidelines do offer choices to the clinician, all of those choices rely on biological conceptions of obesity only. When practitioners encounter a case of morbid obesity that seems to be caused by trauma instead, then they must choose to go outside the guidelines. Kai Brynjer Hagen (see Chap. 10, this book) suggests that the best available evidence in such cases is not captured in guidelines, but rather in the patients' narratives themselves. In cases like these, a clinician must himself evaluate what the best evidence is, in choosing whether to follow the guideline at all.

That is, the complexity of a patient's situation seems to call for a unique approach to their treatment, rather than the application of a general rule. Thus, the ontology of the patient requires that guidelines be treated as advisory rather than regulatory, as a resource rather than a rule. This then calls into question the idea of guidelines as being rule-utilitarian—there will be no 'rule' that benefits more than the singular patient to whom it will absolutely apply (presuming such a patient does in fact exist, which is up for debate if guidelines are written for statistically 'average' patients, for instance, see Anjum, Chap. 2, this book). Even when a guideline represents the treatment paths that tend to work, they will not always present the best thing to do in a particular case. For this, judgement is needed, and familiarity with the patient himself.

> Simply put…
>
> Rule-utilitarianism suggests that we can find a rule that, if everyone follows it, will provide the most utility for the most people, overall. Act-utilitarianism suggests, in contrast, that each individual decision about whether to take a moral action must weigh up the potential consequences and utility of that particular action.

Therefore, rather than rule-utilitarianism, we might accept that act-utilitarianism is the correct way to understand the nature of guidelines: act-utilitarianism allows one to consider the consequences for overall utility of a specific act or decision, rather than of a generally applied rule (for a full review of this argument, see Anjum and Mumford 2017). However, note that this moves the source of the guideline's utility from the guideline itself to the decision about whether to use the guideline: it is not about the nature of guidelines anymore, if we follow this line of reasoning, but about the nature of the clinician's decision in that particular case. Consequently, it is not guidelines but the clinician who would be utilitarian in nature, after all.

There is thus an ontological reason for *not* seeing guidelines as tramlines, even if we do consider them to be utilitarian in nature. We cannot think of guidelines, ontologically, as utilitarian rules that represent (from a public health perspective) the best treatment options for the greatest number of people. In order to bring the most utility to the most people, rather, each case must be considered individually, before it can be decided whether the rule should apply. So the rule is not utilitarian on its own. Rather, it is the act of using or not using the rule that could be utilitarian, if making that decision for that patient also brings about the most possible utility overall. It seems, then, that the best action to take will be the action that brings about the best consequences for that particular patient. Even when we look at it in terms of the most utility for the most patients over all, this will happen only when each individual patient is given the best possible care for that patient specifically (a singularist approach) rather than when all patients who fit into a generalized category are all given the same care (a frequentist approach). But how do we make guidelines specific enough to give the best care to each individual patient, when patients can differ greatly in practice? Before addressing this problem, we will explore the practical reasons for making guidelines more specific than general.

6.3.2 *What Does This Mean for Guidelines in Practice?*

There are practical reasons for not seeing guidelines as tramlines as well; practitioners have good reasons for resisting the imposition of guidelines as rules for their practice. As we saw above, pressure to follow guidelines, and the presumption that they capture the best available evidence, means that practitioners may fear repercussions for not treating them as rules. But, as we see in the obesity case, guidelines do not always capture the best available evidence in relation to the particular patient at hand. So, when we take dispositional causality and person centered medicine as our paradigms for medical science and care, we can see there is a serious conflict between creating a rule to follow in the clinical encounter and using the best evidence available to decide on a course of care. It is simply not possible for a guideline to do both.

Indeed, one of the best known explorations of this problem has been written up by Gabbay and le May (2004), in their introduction to the idea of seeing guidelines as 'mindlines' instead of tramlines. Gabbay and le May show that guidelines play a complementary role to other practices for healthcare practitioners, including referring to other known authorities in their professional networks. That is, guidelines act as additional sources of evidence about best practices, weighed up in relation to practices they already rely upon in deciding what the best available treatment options for their patients are. Practitioners use their own judgment to decide when and if they will incorporate the advice given in guidelines into their practice and decisions. So guidelines are not treated as rules to follow in practice; thus, it does not make sense to develop them or to try to enforce them as such rules.

The best evidence available is going to include evidence related to the particularities of a situation—to the dispositions of the particular patient, and to the specific context of that patient as a person. But this kind of evidence cannot support the formulation of a rule that should be followed beyond this particular case. Once we understand that tramlines simply cannot do the work that guidelines are supposed to do, any conception of guidelines that insists they be followed as rules is wrongly conceived.

6.4 The Epistemology of Guidelines

Guidelines are meant to represent the best options for care in a given situation, given the best evidence available. But this recalls the problem of how we can know which of the available evidence is, in fact, the best. Guidelines developers regularly struggle with this problem. Many of them have adopted a system called 'GRADE'ing the evidence; the principles of the GRADE (the Grading of Recommendations Assessment, Development and Evaluation) working group can be followed in order to rank evidence from best to least, in respect to how reliable it is for grounding guidelines. The claim is that this methodology offers greater flexibility than the original hierarchies of evidence developed by EBM proponents, because it incorporates a wider scope of evidence and then evaluates that evidence via a greater number of parameters. Whereas the hierarchies relied upon the methods by which evidence was generated to rank one kind of evidence higher or lower than another, GRADE methodology takes into consideration perceptions about the reliability of specific research results, sometimes allowing, for example, evidence that was generated by a low-level method on the EBM hierarchy to be ranked higher, and vice versa. For example, if an observational study reported a large enough effect, confidence in its results would be strong enough to rank it alongside an RCT; or, if there were a risk of bias in the way an RCT had been conducted, for instance if not enough women had been included as participants in a drug study, then its results would be ranked with lower confidence, especially in respect to women. However, even this more flexible way of ranking evidence still limits itself to considering the best available evidence to come from organised, clinical trials. This leaves out, for example, mechanistic evidence, which many believe to be of highest importance for determining causality, and patient narratives, key to understanding causality in the single case (see Anjum and Rocca, Chap. 4, this book).

A second epistemological concern in the development and use of guidelines is how to integrate this wider scope of evidence into a single or set of decision-making options for clinicians and patients. Different kinds of evidence require different kinds of expertise and even different kinds of reasoning to be employed in their evaluation. This is the task taken up in recent years, for instance, by the GIN

(Guidelines International Network) AID (Appraising and Including Different) Knowledge working group (Wieringa et al. 2018). One way of tackling this is to engage a diverse group of experts in the developing of guidelines, who are able to evaluate a wider scope of evidence through their joint efforts (Zuiderent-Jerak et al. 2012). A second way is to improve upon the transparency of guidelines, including details about the way in which decisions were made by developers as well as the end results of their deliberations.

6.4.1 Transparency and the Tension Between Flexibility and Standardization

Transparency, that is, is one of the ways that more flexibility is being built into guidelines by developers. When developers are able to be transparent about the choices they have made—for instance, about the reasons for why they made the decisions they did about which evidence was best, and how that evidence translates into the best practices they describe in the guideline—then clinicians have better tools for deciding whether they ought to follow the guideline itself. A clinician might, for instance, disagree with the reasons behind the choices that developers have made, insofar as that clinician would not have made the same choices either generally speaking or in respect to the particular patient at hand.

Of course, there are limits to how much transparency is useful. Too much transparency could mean that guidelines offer no better guidance than the evidence itself to clinicians and patients, who must first decide whether they agree with the experts in order to make use of their guidelines. An effective guideline is developed by experts who can be trusted to evaluate the evidence on the behalf of others who lack either the expertise or the time at the moment when decisions must be made to do that evaluating work themselves. Again, this follows the observations of Gabbay and le May (2004), that practitioners are more likely to accept new evidence when it is promoted as good evidence by trusted colleagues and authorities in the field.

The ontological concerns raised in the previous chapters and section come into play here, as well. Knowing what evidence is best, and knowing what treatment options it is evidence for, requires an ontological judgment about evidence—we have to first have an idea of what good evidence really is, to tell if the evidence we have is also good. Thus, any process by which experts come together to evaluate evidence starts from their assumptions about what makes good evidence. In developing a guideline, it can be supposed that the best evidence available is also the best evidence for the purpose of developing a guideline. If we further see guidelines as rules to be followed, and a good outcome as achieving the best results most often for the most people, then we are evaluating evidence in terms of its quality as a support for a rule. As we say above, this backs us into an uncomfortable corner, in the thick of the tension between flexibility and standardization.

6.4.2 When Should the Particular Be Engaged?

The epistemological factor that needs to be highlighted here is the point at which nuance and particularities become relevant for whether a guideline will be useful. This point, I argue now, is whenever decisions must be made about what are the best available healthcare options for the particular patient at hand. For these decisions, evidence about the patient's condition and evidence about what treatment options are best must come together. Guidelines are meant to offer guidance on how to do this. Often, this point of convergence is assumed to be the point of diagnosis. But, as previous chapters have shown, there is much more to determining which treatments are best than diagnosing the patient. Further, in many cases, no accepted diagnosis fits (see Anjum and Rocca, Chap. 4, this book). And, diagnosis should not present a point when the clinician has to decide which guideline to follow, or whether to follow one at all. Guidelines should be ever-present tools for the clinician, no matter the diagnosis, leading them through the resources available in a way that really helps. So the particularities should be an inherent part of guidelines, acting for the clinician as starting points from which to begin a search, for example.

Additionally, guidance is needed from the moment the clinical encounter begins because, from the very beginning of the encounter, a clinician is observing new and varied evidence about her patient's health, and thus needs correlating knowledge about what treatment options might be available and best. Thus, guidelines ought to offer guidance not only after a diagnosis has been settled upon, but from the moment the clinical encounter begins.

And finally, we consider flexibility to be important in respect not only to how guidelines can be used, but also in how they might be formed. The clinical encounter, that is, should be seen as a resource for gathering further evidence, not only about the individual patient but also about the effectiveness of the guideline itself or the usefulness of the evidence available in respect to that patient and that clinical encounter. New knowledge is gathered at the site of care, and guidelines should not be imposed in a way that constrains such knowledge production, but rather can play a role in enabling it. For instance, as Rocca (2017) recommends, building bridges that enable evidence from the clinical encounter to be taken up by scientific researchers may be the best way to gather mechanistic evidence, which must be observed in the singular case and reported along with its qualitative context. Further, unexpected discoveries frequently occur within the context of the clinical encounter (such as unusual responses to drugs, whether adverse or beneficial), and processes for developing and using guidelines ought to not only allow for this to happen, but also be prepared to learn from serendipity—fortunate, though unexpected discoveries—when it happens (Rocca et al. 2019). Guidelines, in the right form, could build such bridges and enable serendipity.

All of this means that attempts to build in those particularities and nuance by developing any single guideline that will work as a linear and certain pathway through the steps of treatment are misdirected. No matter how wide the scope of evidence, or diverse the expertise and reasoning employed, guidelines developers cannot hope to capture all details that may be needed by a decision-making team of clinician and patient in the clinical encounter. Therefore, it cannot be the function of guidelines to offer a complete set of evaluations and options. Rather, as suggested in Chap. 3, local knowledge, patient narratives and a variety of evidence are needed to determine the propensities that are relevant in this clinical situation (Rocca, Chap. 3, this book). That is, the epistemic role of guidelines cannot be to provide all the knowledge that will be needed in this context (i.e., all the material that will be needed to make the best possible decisions about care).

Let's return again to the idea of Gabbay and le May (2004), that 'mindlines' are created through the tacit and ubiquitous interactions between colleagues and experts within the practice of healthcare, in contrast to guidelines being imposed from outside and accepted as new rules to follow in and of themselves. Guidelines, in their depiction, are useful insofar as they are trusted sources of new evidence. To be trusted sources, they need to be trusted by trustworthy colleagues, for instance; similarly, if they are developed by trustworthy experts, it is the expertise and not the evidence that will be followed by practitioners. So any guideline will have to be transparent, at least, about who made the decisions when creating it, what their claims to expertise in the interpretation and application of such evidence are, and why they made the decisions they made. It is these decisions, then, that are the true content of a guideline (the guidance they offer), and not the actual treatment paths that are recommended. This gives us one way of understanding how guidelines can be formed and transmitted in a useful way—as collections of expert advice, rather than as a simplified rule or set of options.

Consider one of the biggest challenges for guidelines developers—the prevalence of comorbidity among healthcare patients. As we have discussed (see Rocca and Anjum, Chap. 5, this book), the tendency is to divide the medical body (both the body of the patient and the body of medicine's institutions) into disparate parts. But the reality is that many patients present with multiple conditions, which interact in ways that single-disease guidelines may fail to capture.

A related challenge is how to deal with the wealth of potentially relevant particularities, and how does one identify patterns (that is, decide what counts as evidence) in a unique situation? Guidelines must not only allow for intersection, but also for the flexibility that is needed in the face of the variety between patients and the fact that each patient will also change over time. They need to provide the means for telling a causal story about that patient, one that includes emerging causal relations and genuine complexity (see Rocca and Anjum, Chap. 5, this book).

So guidelines must be broad enough in scope to be useful before diagnosis as well as after, and they must be flexible enough to be adaptable to a changing and unique situation. The epistemic role of guidelines, then, is to offer a navigable network of connections between observable evidence in the clinical encounter and generalizable evidence obtained via the results of medical research, as well as information about what treatment options are actually available and likely to be beneficial. That is, rather than a flowchart of pathways to take in a certain direction once the starting point is decided upon, a guideline could present more like a web of data and information that can be searched in any direction and that would highlight interconnections between possible pathways. These pathways ought to come in the form of expert advice on how to integrate the best available evidence into practices that are already in play. Rather than an imposition of a new rule upon practice, then, guidelines would act as a resource for ways to improve upon existing practice, allowing for variation between practices and for the best available evidence to change practice from the bottom up, rather than the top down. This is just one way to imagine how guidelines might work, if we start from a dispositionalist perspective.

To suggest that guidelines should work bottom up, should trace the interaction between expertise and the best available evidence in a way that is both transparent and helpful to practitioners, and should allow for variations within practice and between institutions, however, seems to bring us back to the original problem identified by the proponents of EBM—that is, how do we standardize best practices in healthcare without also linking best practices to authority figures instead of to the evidence? We have argued that the solution is not to create guidelines in a way that makes them rules to follow, or as inflexible as tramlines. In the next section, we explore further how dispositionalism allows us to imagine a middle way, by linking best evidence directly to the individual, and promoting singularism over frequentist or utilitarian approaches to the standardization of best practices.

6.5 Guidelines in the Dispositionalist Way

> In moral philosophy, the idea that right and wrong can be defined by a system of rules has been challenged. If nothing else, it's obvious that two rules could easily come into conflict and one of them has to be sacrificed. Telling the truth is good but not necessarily if a killer asks for the whereabouts of any intended victim. There could be circumstances in which it is right to lie. In response, Dancy [2004] proposed a theory of moral particularism, a view in which each situation has to be understood as complex and requiring its own moral assessment, which could well be unique and unrepeatable. I would favour coupling this with a strongly dispositional version of virtue ethics. Telling the truth tends to be right but not necessarily so. Assessing the whole complex of circumstances might weigh in favour of lying.
>
> Now I think the issue of how to understand and use a guideline clearly relates to this discussion. We could interpret a guideline in a dispositional way rather than as an absolute rule. A particular intervention may tend to relieve a particular symptom but in many contexts it need not be the right intervention to prescribe. If this is right, I think it would be to the benefit of all stakeholders—clinicians, guideline bodies, regulatory authorities, and patients—to understand dispositionalism and particularism. This could be a challenge when rule-based laws and codes of ethics are easy to grasp, but there is a potential benefit to be gained from pushing ahead for a conceptual change.
>
> Stephen Mumford, 'The Notion of Guidelines', CauseHealth blog
> https://causehealthblog.wordpress.com/2016/12/08/the-notion-of-guideline/

6.5.1 So, What Should We Do with Guidelines?

We wrote earlier about the fact that we cannot know about a causal relationship until after it has been observed (see Anjum, Chap. 2, this book). However, even after we have seen the same cause and same effect occur together repeatedly, this does not mean that the causal relationship did not, in that first instance, actually exist. That is, our epistemological state does not determine the ontological case. In the case of guidelines, it seems we are trying to use what we know, epistemically speaking, to say something about what must be true, ontologically speaking. When we set a guideline, we pass a judgment about what treatment options are best, given our evaluation of the evidence available. This judgment is not just a guess, it came about via rational deliberation among experts. However, the best evidence available, as has been shown in previous chapters, still cannot tell us what will actually happen, in a particular case, given the specificity of the mutual manifestation partners involved. In most cases, the best available evidence can only give us a probability range of an outcome occurring; epistemically speaking, probability is often the best knowledge we can have. Thus, guidelines are still useful, even if they cannot prescribe exactly what the best decision will be.

But dispositionalism gives us something further to work with than this. It is true that we cannot use our knowledge to determine exactly what will happen when a particular patient embarks on a particular treatment path. However, as I mentioned in the last section, the clinical encounter itself provides considerable observable evidence about the particularities and nuance needed in deciding what options presented in a guideline to follow. What is important is that guidelines be written in such a way that the general medical knowledge we have already can be met effectively by the knowledge gained during the clinical encounter itself. What the clinician might know to be true about a situation gives her new tools for evaluating the relevance of a guideline for her patient.

What might a guideline look like, then, if we take up the conclusions drawn in this book? Above, we said that they ought to offer a navigable network of connections between observable evidence in the clinical encounter and generalizable evidence obtained via the results of medical research, as well as information about what treatment options are actually available and likely to be beneficial. Mumford, in the quotation above, adds the concerns of particularism and virtue to be considered by those who develop and use guidelines (see also Anjum and Mumford 2017). Add to this what I have brought up in this section, that guidelines provide only half of what is needed for their own effectiveness in the clinical encounter—it is up to the clinician to make observations and the patient to contribute further evidence in order for them to work successfully with the guideline to make decisions. Taking all this together, it could be argued that guidelines are most effective as tools for use in the clinical encounter when they are transparent and accessible in a way that allows patients and clinicians to question the guideline itself where necessary, and to easily draw connections between their particular situation and the more generalizable advice given by the guideline.

In order to be fully dispositional, however, guidelines also need to provide more information about what kinds of mechanisms might be in play, and what kinds of dispositions are likely to affect the treatment path's effectiveness, if chosen. This will require not only a transparent process, but an iterative process, that takes up new evidence gained in clinical encounters and makes it accessible to future users of those guidelines. As in pharmacovigilance, patient reports can play a key role here (Rocca et al. 2019). With the increasing use of data mining techniques and electronic records, paired with clinician and patient narratives when helpful, such information is already taken up in guideline development—it could, perhaps, more prominently shape the way guidelines work, so that using guidelines also means actively engaging with them on a regular basis. The experts whose advice will be accepted by practitioners, as we have seen, comes not only from guidelines developers, but also from colleagues and leaders in the field who have put them into practice and reflected upon their impact. So the development of guidelines (especially if we take them seriously as 'mindlines') does not stop at their implementation, but rather needs to be continued afterward by taking up and disseminating the ways in which individual practitioners have been able to use them in practice. In addition, this kind of new evidence can communicate details about what kinds of patients the guidelines have worked best for, and the contexts in which they have so worked.

In addition, guidelines could offer advice on how to interpret different kinds of evidence that will arise in the clinical encounter, such as what kinds of things to note within a patient narrative that may be clues to relevant dispositions, or other symptoms not immediately evident from a patient's physiology but which may affect treatment options. In this way, guidelines could do more than offer a dispositionalist clinician the right guide to assessing and treating a patient, they could promote a dispositionalist approach to care. Rather than seeing guidelines as presenting ready-made options for a suitably assessed patient, then, guidelines would provide suggestions for how to assess the situation and how to take various psychological and social factors into account when making decisions with patients about their care. Many of them, indeed, are already taking this task on.

Such an approach may seem complicated, but only if we hold on to the utilitarian ideal of finding guidelines that maximise the utility of medical care over a population. Once we have taken a dispositionalist turn, toward the particular patient and the integration of a plurality of evidence types in order to focus on mutual manifestation potentials, then it makes no sense to spend our time developing such rules, if they are not also useful. Rather, it would be more natural to create guidelines that work more like networks and databases, giving access to a constantly updating base of evidence and resulting advice, and highlighting the fact that evidence collection is a continuous process, not ending at the point of diagnosis and treatment choice.

Finally, we at CauseHealth recommend taking an ecological turn in medicine (see Rocca and Anjum, Chap. 5, this book). Guidelines could be part of such a major change in focus, away from specialist focus on different parts or systems of the body, and toward integration with extra-physiological factors. Just as recognizing mutual manifestations, complexity and causal singularism also means that we cannot best study a causal relation in isolation, we cannot expect to fully treat the so-caused condition in a patient by isolating it from its context and environment (see Price, Chap.7, and Low, Chap. 8, this book). Similarly, it is unlikely that a single guideline or set will fully treat that condition without also engaging the interacting factors. Consequently, and importantly, guidelines need to express an acceptance that many clinical decisions will be made in a state of uncertainty. We are unlikely to know, even with the best evidence available, whether a treatment plan will work in the predicted way, because it is unlikely that we will have full knowledge of all the mechanisms, dispositions and, thus, causal relations that may interact with that treatment in producing its effect in that particular patient. The ecological account, that is, takes seriously the complexity of the kinds of interactions that influence an individual's health. Once we take the ecological nature of health and healthcare into account, then, we see that the design of guidelines must also be a holistic process, seeing guidelines as sets of intersecting advice and increasing awareness of possible interactions—guidelines must be living resources, embodying both the uncertainty and the expertise of the clinicians and patients who use them.

6.6 To Sum Up…

One of the key problems in medicine today, as the clinicians we have had the pleasure to work with have told us, is how to handle guidelines. Public management approaches to medicine tend to promote guidelines as rules to follow, and clinicians often feel pressure to follow a guideline even when their judgment cautions them to do otherwise. This 'tramline' approach to guidelines, we have shown, is philosophically as well as practically problematic. Especially when we take dispositions as the ontology of the causal relations that guidelines want to key in on—the best way to cause a recovery, or to counteract the causes of a condition—we see that guidelines cannot and ought not be treated as rules to be followed.

If this is taken seriously, then the development of guidelines must be more than the collective effort of diverse experts on a particular category of disease or subgroup of patients. Rather, guidelines must give clinicians the tools to assess the potential for mutual manifestations between the patient and the treatment options. They must allow clinicians to be flexible with the treatment plan, making changes as the treatment progresses and new evidence arises. And finally, their development does not end with the creation of a rule, but rather continues with the collection of that new evidence, taken back up into the guidelines to provide a continuously improving resource for each new clinical encounter. These features are needed for guidelines to be useful resources for clinicians, and we have used dispositionalism to ground them in the very nature of causation. This chapter, then, has presented one way in which the ontological assumptions we hold about medicine directly affect how our medical institutions work.

References and Further Readings

Anjum RL, Mumford S (2017) A philosophical argument against evidence-based policy. J Eval Clin Pract 23:1045–1050

Anjum RL, Copeland S, Kerry R, Rocca E (2018) The guidelines challenge—philosophy, practice, policy. J Eval Clin Pract 24:1120–1126

Dancy J (2004) Ethics without principles. Oxford University Press, Oxford

Gabbay J, le May A (2004) Evidence based guidelines or collectively constructed "mindlines?" Ethnographic study of knowledge management in primary care. BMJ 329:1–5

Guyatt G, Cairns JA, Churchill D et al (1992) Evidence-based medicine. A new approach to teaching the practice of medicine. JAMA 268:2420–2425

Petursson H, Getz L, Sirudsson JA, Hetlevik I (2009) Current European guidelines for management of arterial hypertension: are they adequate for use in primary care? Modelling study base on the Norwegian HUNT 2 population. BMC Fam Pract 10:70

Rocca E (2017) Bridging the boundaries between scientists and clinicians. Mechanistic hypotheses and patient stories in risk assessment of drugs. J Eval Clin Pract 23:114–120

Rocca E, Copeland S, Edwards IR (2019) Pharmacovigilance as scientific discovery: an argument for trans-disciplinarity. Drug Saf 42:1115–1124

Sackett DO, Rosenberg WM, Gray JAM, Haynes RB, Richardson WS (1996) Evidence based medicine: what it is and what it isn't. BMJ 312:71–72

Wieringa S, Dreesens D, on behalf of the AID Knowledge Working Group of the Guidelines International Network et al (2018) Different knowledge, different styles of reasoning: a challenge for guideline development. BMJ EBM 23:87–91

Zuiderent-Jerak T, Forland F, Macbeth F (2012) Guidelines should reflect all knowledge, not just clinical trials. BMJ 345:36702

Open Access This chapter is licensed under the terms of the Creative Commons Attribution 4.0 International License (http://creativecommons.org/licenses/by/4.0/), which permits use, sharing, adaptation, distribution and reproduction in any medium or format, as long as you give appropriate credit to the original author(s) and the source, provide a link to the Creative Commons license and indicate if changes were made.

The images or other third party material in this chapter are included in the chapter's Creative Commons license, unless indicated otherwise in a credit line to the material. If material is not included in the chapter's Creative Commons license and your intended use is not permitted by statutory regulation or exceeds the permitted use, you will need to obtain permission directly from the copyright holder.

Part II
Application to the Clinic

Chapter 7
The Complexity of Persistent Pain – A Patient's Perspective

Christine Price

7.1 Introduction

For several years I thought my persistent pain story started at the time I experienced a manual handling injury, but now I know it didn't, it started the day I was born. I thought the pain was simply explained by the physical 'damage' in my back, but it isn't, persistent pain is much more complex. Before my injury, like everyone else, I experienced pain periodically, for example when I fell over, when I sprained my ankle or when I burnt myself on a hot pan. I assumed that pain meant I had physically 'hurt' a part of myself and that when it 'healed' the pain would stop. I didn't put much thought into what pain was or what it was affected by. I didn't need to. Following my injury, I travelled a long journey of discovery, learning about the complexities of persistent pain and considering carefully how I could use that learning to better manage my pain. This is an account of that journey.

7.2 The Injury I Haven't Recovered From

My severe pain started on a Saturday morning in July 2008. I'd spent the previous 2 weeks helping to clear out a large Victorian house ready for some improvements. I had been doing a lot of heavy manual handling. I was told later by clinicians that it was this intensive manual handling over a short time period that likely led to my resultant back difficulties.

Shortly after breakfast, my husband and I headed off to a local beauty spot. On the journey out, I could feel my right leg starting to hurt. It was annoying, but nothing too bad, and not enough to stop us going. However, by the time we arrived home

---

C. Price (✉)
Bournemouth, England, UK
e-mail: christine@cprice.me.uk

for lunch I was in a great deal of pain. By early afternoon I was experiencing severe pain in both my back and my leg, and I could barely walk. The pain had progressed quickly. The severe pain continued the next day, and my husband called out a general practitioner. Armed with strong painkillers I assumed that things would settle, however Monday morning saw no improvement, and if anything, the pain was worse. I was totally unable to walk, even as far as the bathroom. I phoned 101 for advice, and they called out an ambulance. I was taken by the ambulance to hospital. My L5-S1 lumbar disc had herniated and it was compressing my S1 nerve root. I was experiencing excruciating back pain and neuropathic pain. I was started on a range of strong painkillers, including morphine.

After 5 days in hospital I was discharged home. For several weeks I struggled to walk more than a few paces and I struggled to sit down or stand. Despite the powerful medication I was still in intense pain. September came around and I was determined to go back to my job as a teacher for the beginning of term. I was lucky, my headteacher was extremely supportive and allowed me to work flexibly.

7.3 Being Treated Within a Narrow View of Pain

I struggled on for many months. I was given epidural injections, very powerful pain medication, and some physiotherapy. Eighteen months after my injury I had back surgery, but unfortunately by then my S1 nerve root had been permanently damaged and surgery did little to relieve my pain. Following advice from a clinician I left the physical demands of teaching and January 2010 saw me start a new part-time self-employed life, which eased my pain situation considerably. I have never looked back.

I was given further injections, further medication and further physiotherapy. Eventually I was told there was little more that could be done for me apart from implanting a spinal cord stimulator. I was duly put on the spinal cord stimulator pathway.

For the first few years most of my discussions with the various clinicians revolved around medications, injections and surgery. I knew no different, and certainly was not adequately equipped to challenge this or move my care into something more. My pain management skills were minimal and centred mainly on managing medications. I was experiencing daily debilitating pain. The 'medical' interventions were not helping me sufficiently. Most of the physiotherapy I received had limited success for me. Although I was being given some basic advice during physiotherapy sessions, I didn't really understand why my body was constantly in pain, what the triggers for my pain were, and what the realistic likely outcomes were for me. I simply didn't understand persistent pain or how I could better manage my pain. I didn't know how to move my pain situation further forward, despite my best endeavours to do so. I was struggling day to day.

7.4 Starting to Learn About the Complexity of Pain

Around 4 years after my injury I was fortunate to receive an episode of care undertaken by Advanced Scope Physiotherapist Matthew Low. It was through this that I started my journey towards understanding persistent pain and how better to manage it.

Matt's approach was entirely different to previous physiotherapists. He focussed on improving my functional abilities and minimising my maladaptive compensation methods, which other physiotherapists had not done. He also started to teach me to understand persistent pain, and in particular *my* persistent pain, and how to manage it better. Matt demonstrated a genuine interest in both my pain story and me as a person. He used sensitive non-judgemental questioning, and carefully prompted me when needed in order to better understand my narrative. Through the sessions he got to know me as an individual and I felt valued. I felt an equal partner in my care, and I trusted him. I felt I was able to disclose anything about my pain, including the irrational fears and emotions it was causing me to experience.

Matt recognised that my understanding of pain, in particular *my* pain, was pretty low. He combined verbal, pictorial and physical explanations with suggestions for reading outside of the physiotherapy sessions in order to improve my basic understanding of pain. He skilfully revisited and extended my understanding throughout the episode of care. I entered this episode of care thinking that persistent pain was basically unidimensional (i.e. there was something physically wrong with my back at that moment in time and it was therefore responding with pain). Matt explained that persistent pain is much more complex and affected by many different things, including what I do during the day and my emotions. This learning was hugely important for me. It opened the door to me understanding what in simple terms might be feeding into my pain, and even more importantly what I could try and do to self-manage my pain. I left that episode of physiotherapy care with a much better understanding of the complexity of pain, and with much better pain management skills.

Following this episode of care, I took a personal decision to stop all medications, and to rely purely on pain management techniques. I withdrew from the spinal cord stimulator pathway. Armed with my new understanding of persistent pain I worked on minimising my stress levels, remaining positive, boosting my resilience and making adaptations to my personal life, work and home. This helped to minimise and manage my pain and I was able to live a much richer, more pain-free and more fulfilled life.

This approach to pain management worked well for several years but when my personal circumstances changed, partly through a change in my self-employment, my pain began to be more problematic for me. I felt I needed some more professional input and so I asked to be re-referred to Matt. Despite the intervening years, Matt and I were able to pick up almost where we left off. The strong therapeutic alliance we had built up together remained.

Since my first episode of care Matt had become an NHS Consultant Physiotherapist, and a key member of CauseHealth. Pain science had moved forward, and predictive processing had become a key focus for research. Whilst Matt's knowledge and skills had further improved, mine had taken a turn for the worse. I remained fully on board with the basic idea that persistent pain is complex and involves more than just the pathology that I presented with. However, during these intervening years I had heard and read various bits and pieces about back conditions, sciatica and pain, which had just served to confuse my understanding of pain, and in particular *my* pain. During this second episode of care, as well as working on the physical difficulties I was presenting with, Matt patiently took me through the basics of understanding pain once more. He also introduced me to causality, dispositionalism and predictive processing (see Low, Chap. 8, this book).

7.5 Learning About Causality and Dispositionalism

Before I could evaluate whether the concepts of causality and dispositionalism would be able to help my understanding, and management, of my persistent pain, I needed to first explore and understand them. In order to aid, and test out, my understanding of causality and dispositionalism, I created the following smallholding analogy.

The analogy is based on a smallholding commune, with individual commune members working together to benefit the whole. It considers how the individual members' dominant traits, or dispositions, dynamically vary, interact and affect the commune, sometimes resulting in a healthy commune and sometimes in an unhealthy, or unwell, commune. The way the smallholding commune works in terms of having a number of dispositions dynamically varying and interacting with one another, resulting in a healthy or unhealthy outcome, compares well to that of an individual person. Every individual person has a unique range and combination of traits, or dispositions, that dynamically vary, interact, and act as causal factors in the health of that person.

7.6 A Smallholding Analogy

Nine young friends decide to club together to buy a smallholding in a beautiful part of Dorset. Everyone will help grow food on the land, care for their animals and look after the smallholding. The aim is to have a vibrant, happy, hardworking, outward looking commune with enough food on the table, and maybe even some to sell. Each of the friends has one particularly strong trait, or disposition:

Carl – catastrophising
Henry – hypervigilance
Amy – anxiety

7 The Complexity of Persistent Pain – A Patient's Perspective

Debra – depression
Sue – sleep difficulties
Patricia – positivity
Rebecca – resilience

Tammy has experienced trauma
Danny has low blood pressure.

Some of these dispositions may have been present from birth, e.g. anxiety, whilst others may have been affected by their upbringing, e.g. resilience. Some dispositions may be related to their physical body, such as low blood pressure.

Each friend's strong traits, or dispositions, vary in intensity over time, often depending on interactions with one another and what else is going on in life for them, including their physical health. Some friends have a stronger individual influence on the rest of the group than others, for example Debra is a strong influence, whilst Rebecca has a much lower influence.

Each hour, each day, each week and each month is different in the commune. One day might include Amy being highly anxious, Debra's depression being minimal, Tammy being troubled with memories of her trauma and Rebecca's resilience being high. On other days there will be a different mix of the friends' dispositional levels.

The following vector diagram (see also Anjum, Chap. 2, this book) illustrates the friends' levels of dispositions at a good time (Fig. 7.1). The dispositions represented by vectors to the left of the centre line are 'negative' and likely to cause the friend's difficulty, whilst those to the right are 'positive'.

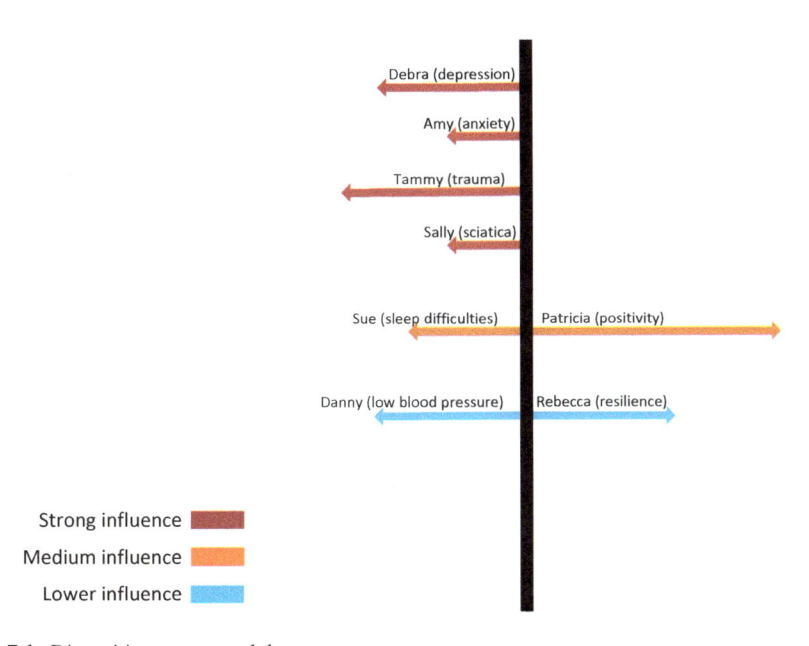

Fig. 7.1 Dispositions on a good day

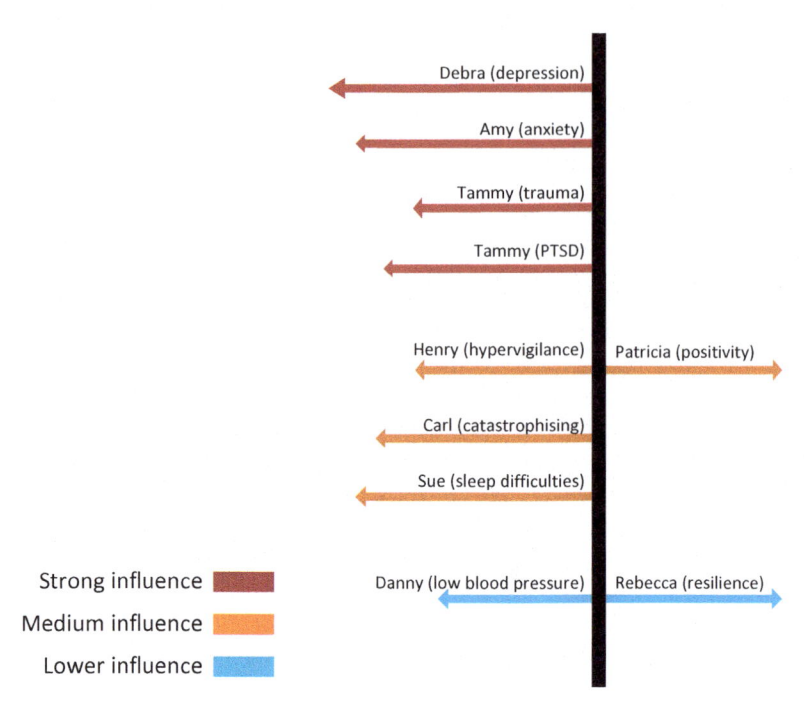

Fig. 7.2 Dispositions on a difficult day

The length of the vectors is an indication of the level of each disposition at that time. The colour of the vectors indicates the strength of their influence. The vector diagram above illustrates the friends' levels of disposition at a more difficult time (Fig. 7.2).

In such a close community, the friends will inevitably have an impact on each other. For example, when Tammy becomes overwhelmed by memories of her trauma and discusses them with Amy, then Amy may find her anxiety becomes worse. Sue may not have had any sleep difficulties for some time but as she listens to Tammy's stories, her tendency, or disposition, to sleep difficulties may become triggered. Patricia's positivity may have a beneficial effect on all the friends. It could also happen that when Tammy and Amy get together the effect of Amy's anxiety on Tammy may mean that Tammy develops a new, second disposition of PTSD (Post Traumatic Stress Disorder). This new disposition of PTSD has 'emerged' from the dispositions of anxiety and trauma.

As well as changes within, and between, the friends themselves, there are other factors affecting the friends. Some of the factors are environmental, for example whether the heating is working or not or whether there is a storm outside. Some may be financial. A bill for repairs to the roof may have come in or some money may have been left to the friends in a will. Some may be social. One of the friend's family may be causing difficulties or the neighbours may be upset by their barking dogs. Some may be biological. Some of the friends may have become unwell with a virus.

All these factors will affect the friends differently. For example, Amy may become anxious over financial concerns and Sue may find it difficult to sleep when there is a storm outside. At such times Rebecca's resilience may reduce.

When everyone is at their best, the heating works, the weather is good, the friends have enough money, their families are supportive, they have no extra physical ailments and they are working well together the 'health' of the commune is good. The commune is 'healthy', and the friends are happy and contented. At this point the group of friends could be considered to have developed a new disposition, that of being a community of mutual support. This mutually supportive community may well provide positive benefits to its members. For example, Carl may benefit from the support of the commune and cease to be hypervigilant in that environment. If Carl moved out of the commune then he may be able to maintain this benefit, for at least a period of time. However, if he then moved into a more dysfunctional community, then this hypervigilance would likely return.

Life isn't always good for the commune, though. As the friends vary, including for example Amy becoming more anxious, Catherine catastrophising and Debra's depression becoming more prominent, then things might start to get a little harder for them. The commune becomes less 'healthy', although outwardly it may still appear to be fine, the friends may still be reasonably content, and the commune may still be functioning.

Unfortunately, at some point in time the commune may cross a 'threshold' and become 'unhealthy' or 'sick'. It is impossible to predict what combination of the friends' levels of dispositions might cause this to happen. The threshold could be crossed due to a combination of Amy experiencing high levels of anxiety, Sue having substantial difficulties sleeping, Debra suffering moderate levels of depression and Rebecca's resilience being unusually low. However, it could be crossed due to a completely different combination. No one friend in isolation is likely to cause the threshold to be crossed, it is the novel mix of their levels of dispositions at that moment in time that causes this.

It may be that the commune oscillates around the threshold, being 'healthy' at times, and being 'unhealthy' and struggling at other times. I find it useful to represent the interactions of the dispositions as a circular vector diagram. The diagram in Fig. 7.3, based on the vector diagram above for a good day (Fig. 7.1), represents the 'health' of the commune on that good day. Each oval is a representation of the strength and level of each friend's disposition. Dispositions within the circle are negative, whilst those outside the circle are positive. The dotted circle is a representation of a dynamic 'threshold' which shows the commune's 'capacity' for health. The higher the level of positive dispositions, the bigger the 'capacity' of the commune to be healthy.

The diagram in Fig. 7.4 is based on the vector diagram for a difficult day (Fig. 7.2) and represents the health of the commune on that particular difficult day. We see that the representation of the dispositions crosses the capacity 'threshold', which is now smaller than it was on a good day.

As the commune becomes 'unhealthy' or 'sick' it starts to struggle. Everything becomes harder and some of the friends' individual traits may worsen. This serves

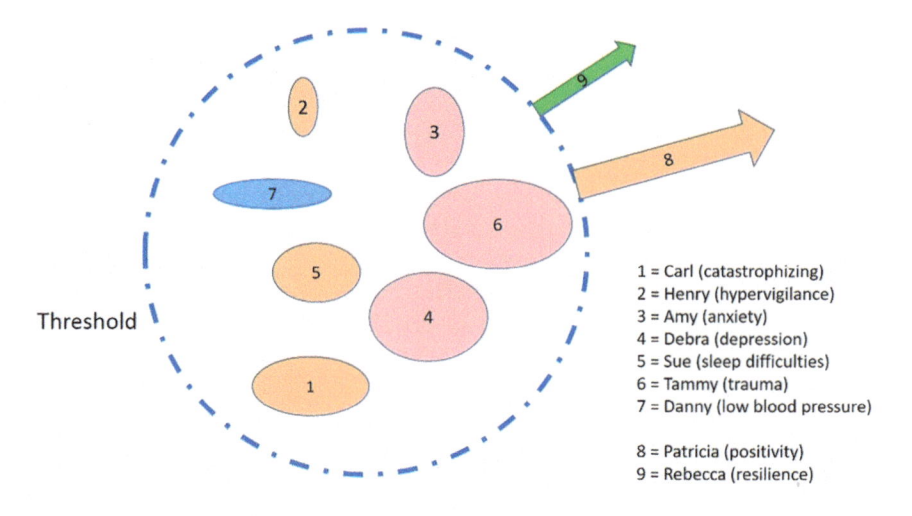

Fig. 7.3 A dynamic threshold for health

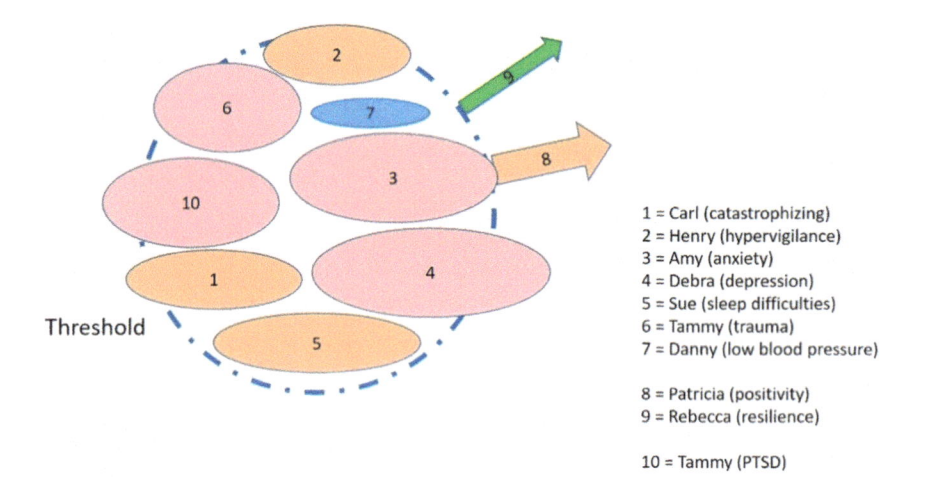

1 = Carl (catastrophizing)
2 = Henry (hypervigilance)
3 = Amy (anxiety)
4 = Debra (depression)
5 = Sue (sleep difficulties)
6 = Tammy (trauma)
7 = Danny (low blood pressure)

8 = Patricia (positivity)
9 = Rebecca (resilience)

Fig. 7.4 Threshold on a difficult day

1 = Carl (catastrophizing)
2 = Henry (hypervigilance)
3 = Amy (anxiety)
4 = Debra (depression)
5 = Sue (sleep difficulties)
6 = Tammy (trauma)
7 = Danny (low blood pressure)

8 = Patricia (positivity)
9 = Rebecca (resilience)

10 = Tammy (PTSD)

to maintain the unhealthy situation, maybe even making it worse. The circle, or threshold, is smaller this time, in line with the levels of the positive dispositions being smaller, and so the 'capacity' of the commune to be healthy is reduced.

Once the commune has crossed over the threshold and become 'unhealthy' or 'sick' then it can be difficult to improve the situation and get it back over the threshold to become a healthy commune once more. There is unlikely to be a quick 'fix'. There is unlikely to be a single 'cause' that can be addressed in isolation in order to fix the problem. For example, just a clinician supporting Debra to improve her depression may not have enough of an impact on the overall complicated mix of the friends' dispositional levels and therefore the commune. It may also not be effective without addressing, for example, the commune's financial difficulties at the same time.

Improving just one friend's dispositional level is unlikely to improve the mix enough to bring the commune back over the threshold. Attention needs to be paid to the multiple factors, or causes, including both the friends' dispositional levels and the external factors, such as finances, social relationships etc. that impact on them. Professional input could be helpful in addressing the variety of factors. For example, support in resolving financial difficulties, support with depression or support with any medical conditions such as low blood pressure. Advice could be given to help support the commune self-manage their situation more effectively. As well as trying to reduce the levels of the more negative dispositions, such as Debra's depression and Amy's anxiety, then boosting the more positive dispositions could help the group. For example, boosting Patricia's positivity might help 'lift' the group. Supporting all the friends, in a variety of different ways, will give the best chance for the group to rise back above the threshold and become a healthy, happy, productive commune once more.

7.7 The Analogy Explained

We all have a unique set of traits, or dispositions. An individual's dominant traits, or dispositions, might be, for example, anxiety, hypervigilance, catastrophising, depression, sleep difficulties, positivity and resilience. That individual may also have experienced trauma and have low blood pressure. These dispositions will vary and interact, as they did in the commune.

In the same way these dispositions were affected in the smallholding commune by external factors, such as unexpected bills or social problems, they will be affected within an individual. For example, an impending court case might cause an individual's anxiety levels to rise, or a break up with a partner might cause depressive levels to rise.

Sometimes, a specific mix of the relative strengths, weaknesses and interplay of these dispositions may cause that individual to cross a 'threshold', which is unique to that individual, and they may become unwell. The individual's unique genetic makeup, existing health and health dispositions will be factors in the presentation of their illness. For example, at any one time, one individual may be pre-disposed to suffering Chronic Fatigue Syndrome, whilst another may be pre-disposed to suffering a chronic pain condition, or generalised anxiety disorder or maybe shingles.

This doesn't of course mean that individuals who are happy, stress free and who have few outside pressures do not become unwell. One of their dispositions, or causal factors, may be physical, for example in my case I have a damaged S1 nerve root. This factor may be very dominant, either in short bursts, or for extensive periods, so causing the individual to cross a threshold, in my case of pain, for either a short or long period of time. Although this factor may be dominant, all the factors still vary and interact together, giving a varying experience of pain and wellness.

In terms of recovery, or improvement, in the same way that the causal factors associated with the smallholding commune becoming unhealthy are complex, so it is with the individual. Trying to address one factor alone is unlikely to improve an individual's situation enough to allow them to cross back over the threshold and return to good, or better, health. A wide-ranging approach is needed. This might include the need for mental health support, GP support, counselling, peer group support and social support.

Crossing back over that threshold and returning to good, or better, health is not always easy. Perhaps the most important first step for that individual is to understand that there are many causal factors involved, which dynamically vary and interact with each other. These causal factors ideally need to be supported and addressed together.

7.8 Combining Causality, Dispositionalism and Predictive Processing

I currently mainly suffer from neuropathic pain, caused by 'damage', and ongoing irritation, to my S1 nerve root. I wanted to know how the nerve signals being generated from this nerve root, often spuriously, might be processed in me as pain and so, with Matt's help, I sought to understand the predictive processing model of pain. Fundamentally the predictive processing model of pain considers peripheral sensitisation, and looks at how anxiety, emotion, expectation and attention may change and impact pain. With the help of Matt, I was able to combine the basic concepts of causality, dispositionalism and predictive processing to come up with a simple understanding of MY pain that works well for me.

7.9 A Simple Understanding of My Pain

I have in my mind/body a 'model' (predictive model) which informs me as to whether to give an experience of pain or not, in a variety of circumstances, based on presenting factors. When a part of my body, in this case my damaged S1 nerve root (which may be being irritated by, for example, position, load or temperature), emits an 'impulse', then my predictive model considers this factor, along with other factors, to evaluate whether to give an experience of pain or not. These factors include my current novel mix of the levels of my traits.

I have a number of personal traits, or dispositions, which vary over time. For example, I have a tendency, or disposition, towards anxiety and poor sleep. I am naturally positive and have high resilience. I experience interaction between these dispositions. For example, my sleep is likely to be worse when I am anxious, and my resilience is likely to be reduced when I am sleep deprived. Some of these dispositions have a stronger influence than others on my presentation, for example anxiety and poor sleep have a greater impact on me than positivity. I am also affected by external factors. For example, my anxiety will increase if I experience work place bullying or an unexpected household bill, and my positivity will increase whilst experiencing success. I have an ever-changing novel mix of the levels of my dispositions. At a 'good time', my anxiety might be low, my positivity high and I might have had good sleep. At a 'bad time', my anxiety may be high, my resilience low and my sleep poor.

My predictive model 'knows' what combination of dispositional levels and other factors, including the impulse from my S1 nerve root, are likely to be 'ok' and don't need a response of pain. If the combination of factors at a moment in time, including the 'impulse' from my S1 nerve root (which is likely for me to be a dominant factor), matches the predictive model of being 'ok', then no action is taken, and no pain emerges. If not, then pain emerges to alert me to do something to stop the irritation on the S1 nerve root continuing. Changes in the novel mix of my dispositional levels, and my S1 nerve root impulse, may, or may not, be sufficient to change whether I experience pain or not.

My experiences inform my predictive model. These experiences might result in the predictive model being changed. In order to improve my pain situation, then I would need to work on optimising my personal factors, e.g. anxiety, sleep, resilience and positivity, my physical factors, e.g. S1 nerve root irritation and also external factors, e.g. temperature, finances and work conditions. This is because my predictive model takes the combination of these factors into account when deciding whether to give me a pain experience following an impulse from my S1 nerve root. Improving one factor only is unlikely to bring about sufficient change.

7.10 How Has Understanding Pain in This Way Helped Me?

Understanding pain has been hugely important to me in terms of my learning to manage my pain situation and lead as full and fulfilled a life as possible. Having a narrative in my head about what is likely affecting my pain and what is happening with my body, which I can fully identify with, is very important to me. It provides me with the confidence to try different things, cope when I am experiencing a flare of symptoms, work out different ways of doing things in order not to 'wind up' my condition and generally live well despite living in pain.

Although important, on its own this understanding isn't enough to successfully manage my pain condition on a day to day basis. I need to use this understanding to seek ways to manage and improve my condition and continue to live as well as possible.

In overall terms I view my pain management situation as needing to reduce the physical irritants on my S1 nerve root, and to optimise my personal, social, health and emotional factors. I also need to consider, and hopefully address, any negative thoughts, beliefs and past experiences which might be acting as factors in my pain

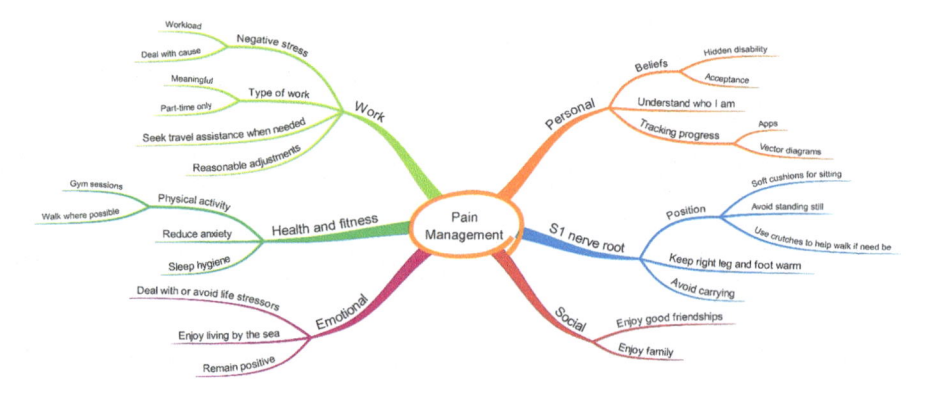

Fig. 7.5 Mind map illustrating the complexity of my pain

experience. As part of this I need to recognise, and hopefully improve, any negative dispositions I have, such as anxiety, and build on my positive dispositions, such as my resilience and having a naturally positive outlook.

I created a mind map (Fig. 7.5) to highlight the areas I need to attend to if I want to live as pain free a life as possible.

The mind map is complex, reflecting the complexity of pain. There are some S1 nerve root specific elements, for example using soft cushions when sitting, keeping my right leg warm etc. There are also elements that are more general, for example good sleep hygiene, enjoying good friendships and reducing anxiety.

It is hard to address all these factors at the same time when trying to improve pain, and it is hard to work out whether the measures you are taking are having a positive impact on your pain, or not. In order to provide a focus for improvement, and to track my improvement, I decided to use the idea of vector diagrams. The vector diagram in Fig. 7.6 shows factors I identified as being the main contributors and improvers to my pain experience in January 2018. The length of the vector gives an indication of the 'size' of the factor, as I saw it at the time. The red vector at the bottom gives an overall indication of the 'size' of all the pain contributors combined, and the green vector at the bottom an overall indication of the 'size' of all the pain improvers combined. The aim is to decrease the pain contributors and increase the pain improvers as much as possible.

The diagram in Fig. 7.7 shows how I rated the same factors in August 2018. Notice that there are some elements I have been able to change, but some that have stayed static. More work to do!

Using vector diagrams works well for me. They give me an instant visual indicator of the problems I have chosen to work on, and I can see visually whether I am making progress with those elements or not, and how much progress.

In my experience clinicians put more emphasis on negative pain factors, the pain contributors, but I think it is helpful to put an equal emphasis on positive factors, the pain improvers. When you are in pain it is far too easy to focus on the negative. I find putting an equal focus on positives really helps to take control of my pain situation and move it forward.

7 The Complexity of Persistent Pain – A Patient's Perspective   125

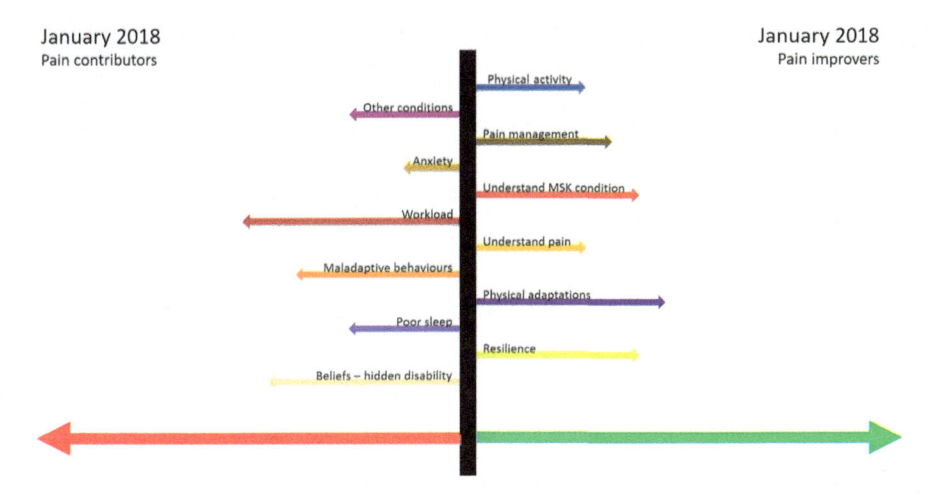

Fig. 7.6 The main contributors and improvers of my pain in January

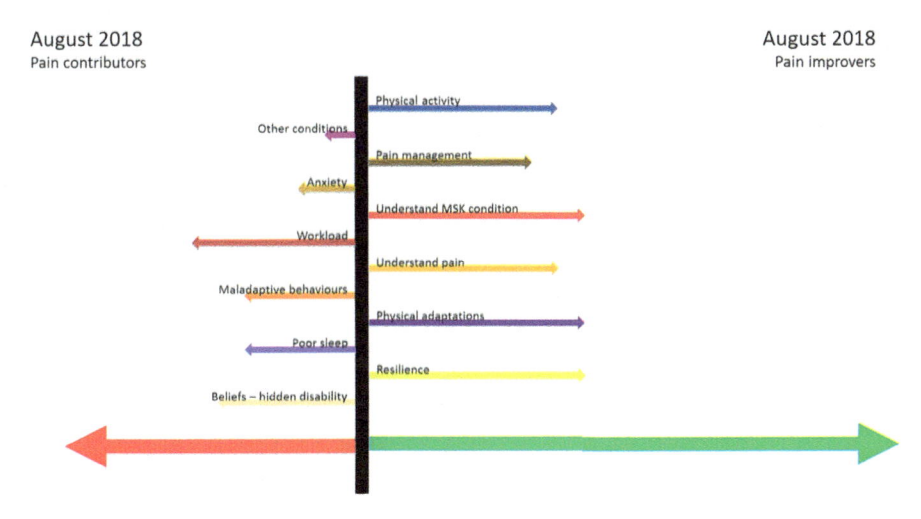

Fig. 7.7 The main contributors and improvers of my pain in August

7.11 The Complexity of Persistent Pain

During the first few years following my manual handling injury I was treated predominantly, if not solely, within a 'medical' model. I had no understanding that persistent pain was complex, and looked only to physiological factors, medications and surgical interventions. I struggled badly with the pain I was experiencing and had few pain management skills.

Following interventions by physiotherapist Matthew Low I was introduced to an understanding that persistent pain is inherently complex. Through my understanding of causality, dispositionalism and the predictive processing model of pain, and with Matt's help, I have been able to come to a much better understanding of pain, and in particular *my* pain. Understanding that persistent pain is inherently complex has enabled me to develop my pain management skills, focussing them on a wide variety of causal factors.

Good understanding and good pain management has enabled me to lead a much richer, more pain free and more fulfilled life.

Open Access This chapter is licensed under the terms of the Creative Commons Attribution 4.0 International License (http://creativecommons.org/licenses/by/4.0/), which permits use, sharing, adaptation, distribution and reproduction in any medium or format, as long as you give appropriate credit to the original author(s) and the source, provide a link to the Creative Commons license and indicate if changes were made.

The images or other third party material in this chapter are included in the chapter's Creative Commons license, unless indicated otherwise in a credit line to the material. If material is not included in the chapter's Creative Commons license and your intended use is not permitted by statutory regulation or exceeds the permitted use, you will need to obtain permission directly from the copyright holder.

Chapter 8
Above and Beyond Statistical Evidence. Why Stories Matter for Clinical Decisions and Shared Decision Making

Matthew Low

8.1 Musculoskeletal Disability

Musculoskeletal (MSK) disability has an enormous effect on the quality of life of millions of people worldwide. In the UK alone:

- One in four, or around 9.6 million UK adults, many of working age, are affected by MSK disability.
- 30% of GP consultations in England are MSK related (Department of Health 2006).
- 10.8 million working days are lost per year due to MSK conditions. A large number of co-morbidities, including diabetes, depression and obesity are associated with MSK conditions (Arthritis Research UK 2013).
- In 2013, more than 25% of all surgical interventions in the National Health Service are MSK related and this is set to rise over the following 10 years (Arthritis Research UK 2013).
- In 2012, £4.76 billion of National Health Service spending each year is on MSK conditions (Department of Health 2012)

I work as a Physiotherapist in the UK, helping people with MSK disability. Patients are often referred to me because they have not improved with previous treatment. I receive requests from diverse other healthcare professionals such as Orthopaedic Surgeons, Rheumatologists, General Practitioners, Geriatricians, Neurologists, Psychologists as well as other Physiotherapists and Occupational Therapists. MSK conditions can be treated in a number of ways, using diverse treatment approaches, in varying sequences and concentrating on different aspects of a presentation. Some conditions may improve by focussing on lifestyle factors, others by addressing physical or psychological factors.

M. Low (✉)
The Royal Bournemouth and Christchurch Foundation NHS Trust, Christchurch, Dorset, UK
e-mail: Matthew.Low@rbch.nhs.uk

Even these categories of influencing factors may be misnomers, due to the interrelations among them. Exercise, for example, may be described as a physical intervention, but has clear psychological effects and requires confidence and motivation to perform. Exercise also exists within a social situation, environment and circumstance. Physiotherapists are privileged in that they, within an accepted social context, have permission to touch, hold and physically assist people through movement in both assessment and treatment. This is a tacit form of communication as well as a method of therapeutic interaction and may assist with the re-education of movement, perhaps by building confidence or trust in one's body through physical reassurance. It is for these reasons that Physiotherapists and other Physical Therapists commonly see people with MSK conditions.

The field of MSK has always interested me; like a detective, who would try to investigate and scrutinise all the evidence to catch the criminal, an MSK practitioner needs to unpack the complexity of a person and their symptoms to get to the heart of the matter. When first learning about being an MSK practitioner, it seemed to me that if you knew all of the research data, or had all the knowledge of the pathology, the rest would just fit into place. Very quickly, however, I found out that this is not the case. There are many skills and types of knowledge—particularly about people—that are required. It is not all about knowing the numbers.

The majority of the patients I see present with their main symptom being pain. Pain has been described by some in practice and research in an objective and impersonal way, such as on a scale between 0 and 10. This may suit quantitative research designs but in no way shape or form does this describe how pain affects a person. Every person has individual hopes, thoughts, feelings, aspirations, abilities, relationships and biographies. The idea of trying to assign a numeric value to these very individual and personal aspects of life seems to me to be both counter intuitive and destructive. In fact, it is an unreasonable expectation that one could examine each aspect independently without regard to the whole person—they play a role not only in a person's life, but in relation to how they experience their pain. I always focus on a person's pain experience.

A person's pain experience usually has a significant impact on what a person can or cannot do. They lose their agency, which is a person's ability to participate in the everyday things that they want to do, such as the normal activities of daily living, but also in things that are meaningful to them, such as family activities, sporting activities and so on.

The patients I tend to see also suffer with persistent pain. By that I mean that the pain has lasted longer than would normally be expected—it has not resolved within an expected time frame. Most patients attend their appointments with me without providing a specific history of injury or trauma, and I can see that their symptoms have gradually worsened over a period of time, often without a clear reason. Some of the problems with pain are that it manifests in a very individual and personal way, it affects relationships, and it changes depending on the situation and circumstances of the environment that a person lives in. Pain is inherently context dependent. By that I mean, amongst a number of things, that a person's history and current situation can change the experience of pain. For instance, if a person is running away from a potentially life threatening situation, they may not feel their ankle break when they twist it, but after experiencing something like that, they might at another time feel severe pain when gently rolling their ankle as they walk over uneven ground.

The non-specific nature of persistent pain presents a number of challenges, particularly if one assumes a biomedical perspective. A biomedical viewpoint is a traditional and very successful form of medicine that assumes a linear interaction between cause(s) and effect(s). An example of this would be identifying a bacterium that causes a disease (such as tuberculosis) and treating it by using antibiotics (medications that affect a bacterium's cell wall or change its genetic makeup) that then eradicates the bacterium and consequently the disease, and thus the symptoms are cured. In MSK conditions, however, the cause(s) are usually multi-dimensional and are inter-connected with other causes, which interact in ways that are not predictable. For example, in low back pain, a person may suffer with severe symptoms of pain and disability and the tests (such as x-ray, magnetic resonance imaging and blood tests) are unable to identify a specific cause and the diagnosis is therefore typically labelled as non-specific low back pain.

The research literature suggests that the majority of cases of low back pain in primary care are non-specific (Koes et al. 2006). There are people who present with specific causes of back pain, sometimes serious, but their symptoms are usually very atypical in that they affect a person's systemic health, leading to weight loss, malaise, fever, or night pain with sweats, none of which are in keeping with an MSK disorder. These presentations are rare and the medical literature suggests that they constitute 1–2% of cases (Koes et al. 2006). Therefore, in addition to the challenges presented by non-specific back pain, such as a lack of an easily identifiable causal relationship to focus on treatment, even those with specific diagnoses present with the challenge of having a distinct and personal symptom profile.

In order to make sound and clinically reasonable decisions for the treatment of MSK disorders, a practitioner needs to take these challenges into account. Therefore, a coherent strategy based upon the judicious collection of pertinent information and evidence that is centred on the patient is of paramount importance. Understanding the clinically relevant and biographical context is needed to begin to unpack the clinical picture and the only valid source for this evidence comes from the clinicians and patients themselves. However, many healthcare scientists and practitioners may not trust the information provided about a condition or treatment from a clinician or patient as it may present a number of confounding factors, including bias, based on personal experience or anecdotal information. Such untrustworthy information may cloud judgement and lead an optimal therapeutic strategy astray. To overcome this, a movement of evidence based medicine (EBM), or evidence based healthcare (EBHC), came about in the early 1990's.

8.2 Evidence Based Healthcare: The Heart Is in the Right Place, But...

EBM is explicitly cautious about justifications for treatments that are not grounded in 'trustworthy evidence'. The EBM and EBHC framework states that degrees of trustworthiness can be described in a hierarchy of evidence, in which research evidence generated by controlled trials is valued higher than the evidence generated by uncontrolled trials. In short, the EBHC movement acknowledges "the role of all empirical observations", however, "it contends that controlled clinical observations provide more trustworthy evidence than do uncontrolled observations, biological experiments, or individual clinician's experiences" (Djulbegovic and Guyatt 2017).

This is due to the three main principles that EBM is based upon, according to Djulbegovic and Guyatt (2017). Firstly, "not all evidence is created equal, and… the practice of medicine should be based on the best available evidence." Essentially, those observations drawn from any uncontrolled methodology such as clinical expertise are ranked as the lowest form of evidence, due to risk of bias, and randomised controlled trials (RCT) are ranked as the highest form of evidence.

Secondly, EBM represents "the philosophical view that the pursuit of truth is best accomplished by evaluating the totality of the evidence, and not selecting evidence that favours a particular claim." Consider that co-morbidities are usually excluded from high quality research trials and therefore are not reflective of the reality of clinical practice. That is, for a study population, a treatment may seem to have been 'effective' for a particular condition, but in practice, for the same treatment the presence of co-morbidities may cause harm. Paradoxically, it is the judgement of the clinician that is required to evaluate and synthesise the totality of the available evidence within the context of the individual patient and their environment, including taking into account various co-morbidities that exist in the clinical setting. Yet it is this clinical judgement that is felt to be untrustworthy in the first place.

Efforts are made to reconcile the paradox in the third and last principle: "Evidence is, however, necessary but not sufficient for effective decision making, which has to address the consequences of importance to the decision maker within the given environment and context. Thus, the third epistemological principle of EBM is that clinical decision making requires consideration of patients' values and preferences." But, this appears to 'bolt on' patient preferences to the end of the aforementioned principles and then offer no insight on how to gather the complexity of evidence together in a person centred way.

One of the natures of 'truth' that the EBM/EBHC movement implicitly suggests, is that statistics, given the 'objective' status of statistical methods, have a higher 'truth' value than other sources of evidence, such as subjective or inter-subjective accounts (e.g. through the sharing of language, experience and understanding) (Øberg et al. 2015). The analyses of studies are often communicated to patients in an absolute or deterministic way, usually in the form of data such as the statistical significance of a finding, the size of a treatment effect, the confidence interval, or the probability of outcome. This assumes a specific philosophical view of what causality is, namely a Humean view of causality (see Anjum, Chap. 2, and Anjum and Rocca, Chap. 4, this book). On the one hand, the clinician may feel confident that the information is 'factually correct', but on the other hand the clinician is assuming that the information that they are providing to the patient recognises all eventualities and all contexts. However, in reality they are giving information based on a 'closed world' where many factors have been controlled for: a world distinctly separate from that of the open, complex and context sensitive world that the patient actually exists within.

I am not suggesting that epidemiological data or carefully controlled RCT's are not helpful, in fact they can be seen as the best available way to understand an average treatment effect on an average person from a clinical trial. This could act as a baseline measure of how effective a treatment may be, but RCT's need to be seen in full light of the limitations that they have, rather than being put up on a pedestal where they are interpreted as the facts of the matter that determine clinical decision making.

8.3 Therapeutic Alliance: A Dispositional View

EBM/EBHC has been described as a map that creates guidelines for treatment or intervention. As has been shown above, however, EBM/EBHC may serve as a map but it does not describe the terrain. The treatment effects of a person who has such individuality, with a complex, sensitive and biographically inscribed biology, simply cannot be reduced to a sequence of recurring events that are seen to occur time and time again. In other words, the Humean view of causality will not do. A dispositional view of causality, however, has the ability to change the perspective of both the clinician and the patient, moving them toward a much clearer perspective that embraces such complexity.

To recap from Part I of this book, causes are dispositions or 'powers' that may tend toward an effect rather than necessitate one. They can operate amongst other dispositions and may lie unmanifested but still be present—they may, for example, still be acting in a way that creates a stable situation, but in such a way that no observable effect is seen. In addition, they interact in non-linear ways and depending on context. This, in my mind, makes absolute sense and accounts for the apparent discrepancies of observed events, but also forms a distinctive and clinically relevant framework for helping people (Low 2017).

Rather than assessing a person, establishing a diagnosis and then looking to the research evidence to inform us how to treat a patient with their preferences, our attention should be particularly drawn towards the person's story. A person's story is a phenomenological account that far supersedes that of a descriptive and categorical diagnosis with regards to how to frame and personalise a treatment or management approach. A categorical diagnosis may provide clinical options for treatment but they are often shaped within broad guidelines. An example of this would be 'education, exercise and weight loss' for the management of hip osteoarthritis. A person's story establishes a unique biography that sets up the clinical context by opening a window of insight into a person's biology. For example, a person's challenging childhood may shed light on their behaviour towards comfort eating and the complex relationships they may have towards those who may be able to support them. What then lies beneath the statistical data is a person who has hopes, fears, relationship and lifestyle considerations, as well as a rich number of historical events that have influenced them.

The understanding of a person's story can elicit copious amounts of causal elements that are essential, not only for the clinician but for the patient, in understanding and coming to terms with their pain experience. These causal elements, or dispositions, can be inferred through the creation of a co-constructed narrative. Both the patient and the clinician bring their understanding together within a space where they each attempt to make sense of the situation. This sense-making is inter-subjective; the clinician is interpreting and coming to terms with what the patient is explaining, demonstrating and exploring while in exactly the same way, the patient is trying to make sense of the clinician's thoughts, feelings and perspectives. I believe that recognising that this sense-making process, that is bi-directional between therapist and patient, is essential to building a strong, therapeutic alliance. There is emerging research evidence to support the idea that having a strong therapeutic alliance improves pain outcomes in patients with persistent pain. Factors such as trust, open communication, using a whole person approach, tailoring an individualised plan and the ability to work through challenges in the patient-clinician relationship have all been identified as enhancing this alliance (Kinney et al. 2018; Bunzli et al. 2016).

Often, in graduate and postgraduate training, a clinician is asked to divide a clinical consultation into subjective and objective phenomena. The 'subjective' information is the information that the patient gives to the clinician, including their story and the 'objective' information is the clinician's observations, measurements and judgements. This way of understanding 'subjective' and 'objective' phenomena undermines the patient as it assumes that the clinician's observations and judgement are objective whilst the thoughts, beliefs and perspectives of the patient are 'subjective' and therefore questionable. It is therefore paradoxical that clinicians who claim that they are 'evidence based' appear to pay only lip service to the idea that 'patient preferences' are on an equal footing with clinical research and clinician judgement when making clinical decisions.

Indeed, a careful physical examination that takes into account the context of the story is essential in gaining trust with the patient. The sensitivity, order and coherence of the physical examination unveil more than positive or negative tests that fulfil diagnostic criteria. The examination may reveal the distress, the inability to cope and painful memories of living with pain. This clinical information that provides clinical context is essential in formulating a person centred management plan. One way to bring this together is by using a mind map where a person's story, physical examination findings, clinical investigations and judgements of the clinician form this co-constructed narrative. This bringing together of the normative and the narrative (Launer 2018) seems to me to be far more humanising than the division of a person into constituent parts: biological, psychological and social elements that are further separated into subjective and objective data findings.

I use a mind map, where all the relevant information from the patient is collected, honoured and placed in a timeline. This allows the patient to make sense of all the available information in one place. The patient has explicit permission to give feedback and change any aspect of the mind map if they feel that it does not capture their reality and the situation. The mind map is personal to the patient, so it is important that the patient has the autonomy to change the mind map to reflect their experience. Following the co-construction of the mind map, a vector model can be made, conveying the complexity of interaction between causal elements (see Price, Chap. 7, this book). It is the use of the vector model that allows a patient to see that the variability and context-dependent nature of their pain experience could be better understood (Low 2017, 2018). Causal elements may be non-linear, interactive, contextual and time sensitive. Understanding this can help a patient look beyond the false understanding of their pain being caused by single elements that stack up or aggregate. Assumptions that causal relationships are linear can, for example, result in surprise when pain is experienced even when patients purposely choose not to do activities or tasks felt to increase their pain.

Instead of using statistical data to determine the prognosis and management of a patient, a dispositional view allows and even encourages the context sensitive use of propensities that have been carefully considered in light of the individuality of the patient. Instead of only describing the probability based upon a research population, the individuality of the patient is also considered within the context of how that person relates to the research population.

The use of terms such as 'propensity' or 'tendency' avoids the deterministic qualities, like inevitability, that can come with MSK diagnoses such as osteoarthritis (OA). Although the incidence of OA increases as we get older, it manifests with a significant amount of variation in different people with different experiences. Socially, it is believed that the presence of osteoarthritis always causes pain and disability, but this is not the case. People believe that a diagnosis of osteoarthritis will mean that their joint will erode, crumble or fall apart (Barker et al. 2009) leading to unnecessary fear, anxiety and loss of activity and agency. This self-fulfilling cycle perpetuates or leads to more pain and disability. However, osteoarthritis is a complex process that is affected by multiple systems including the immune, endocrine, and neuromusculoskeletal systems. Amongst these systems, the experience of pain is modulated by multiple social, psychological and biographical factors, some of which can be tempered, altered and positively changed. Reductions in pain and improvement in function is possible and the manifestation of pain and progressive disability is most certainly not inevitable. Using a mind map and vector model of a person's individual situation can make a difference and turn the tide.

The phrase, "treat the person, not the scan", commonly used in medicine, cannot be stated enough as one person may have significant OA changes seen via imaging and present with joint stiffness but no pain, whereas another person with very mild OA changes may present with disabling and distressing pain. This is true across an entire spectrum of MSK diagnoses, including the most commonly found conditions such as low back pain, fibromyalgia and neck pain. Often, MSK diagnoses do not exist in isolation. Mind-body dualism, what the philosopher René Descartes is known for, underpins biomedicine and is reflected within the objectivity of EBM/EBHC in the way that distinct quantitative physical and psychological measurements are used to draw causal inferences without considering the interaction and non-linearity of causal elements that exist within a complex situation such as the lived experience of a person. A dispositional perspective, in contrast, can ground a non-judgemental viewpoint of a person's lived experience through the use of a mind map and a vector model of powers. Aspects of the clinical picture can be displayed to convey the complex nature of a situation but still allow space to develop a management strategy that a person can understand, accept and initiate.

8.4 Bringing the Totality of Evidence Together

When implementing a person centred clinical reasoning perspective, how should one proceed? There are many different perspectives on this; my personal approach is to centre the process on the patient. Their story, viewpoint, narrative and situation are central and of the utmost importance as they provide the vital context of the therapeutic encounter. The evidence drawn from clinical research is then applied within the context of the patient, and with sound judgement the clinician then evaluates the applicability of the research to the patient, at this time. Then, both the patient and the clinician bring together the narrative and normative information alongside the clinical research in order to create a truly shared perspective on treatment and management using the mind map and vector model.

All interventions in healthcare aim to improve a patient's healthcare need, situation or circumstance and they carry a form of risk that may delay progress or adversely affect the outcome. In MSK disorders, the most common risk to a patient in a physical therapy clinic is the potential to worsen a patient's pain experience. Therefore, careful discussions of the potential risks and benefits of any future action should be made prior to the implementation of treatment. The use of the mind map and vector model makes this process easier and better understood, allowing treatment recommendations to be made in a coherent fashion. The added benefit of doing this allows the clinician to provide options if symptoms adversely change and offer potential reasons why this may happen. For example, a patient with persistent non-specific back pain may present with fear of bending their back, as they may believe that their symptoms are indicative of the spine being damaged. The reason for the patient's fear may relate to their previous experience of bending to lift an object in the past, for example, or by well-meaning family, friends or healthcare practitioners telling the patient that the pain when bending is harming their spine. The treatment option that may be the most appropriate for this situation is a graded exposure form of treatment whereby the patient starts to understand how their fears relate to their pain and are encouraged to participate in the activities that they fear in a graduated way. It is important to understand why the patient is fearful and this can be explored narratively and by performing behavioural experiments or safe versions of the feared activity. This creates an opportunity to understand the cause and relationship of fear to the current situation and is made explicit by using the mind map. The mind map can help explain a vicious cycle of fear, avoidance of activity and deconditioning that is perpetuating and maintaining the pain experience. The treatment option of graded exposure may increase the patient's symptoms, so a strategy of the patient being comfortable with the treatment and having options of what to do if their symptoms increase is therefore important.

The details of this are highly context dependent and cannot be demonstrated through statistical data, nor can the clinical adjustments and variations that allow a patient to stay in control of their symptoms. Self-efficacy or the sense of a person's ability to feel in control of their circumstances or situation is, amongst a number of factors, important for health outcomes (Holden 1992; Keedy et al. 2014). Using the mind map and vector model are useful tools to convey the totality of evidence in describing the patient's story, their context, the clinical research evidence, the social situation, the risks and benefits of treatment options as well as opportunities to tailor a personalised management plan for the patient.

I have previously written on related cases, of a person with medically unexplained neck, thoracic and shoulder pain (Low 2017), and of a person with persistent and recurrent low back pain (Low 2018), which more fully explain this approach. The former describes a little more of a philosophical account whereas the latter describes a clinical account in greater detail.

In summary, understanding the clinical research that is epitomised traditionally through statistical data does not help the clinical encounter. If anything, there is a greater risk of alienating and depersonalising the lived experience of the patient. Respecting the patient, rather than the statistics, at the centre of the evidence brings with it key aspects of a therapeutic alliance, characterised by connectedness (Babatunde et al. 2017). This connectedness, established through the creation of a co-constructed shared narrative, improves the therapeutic alliance, adherence to treatment from the patient (Babatunde et al. 2017) and personal job satisfaction for the clinician. Although biomedicine has brought about huge successes in the eradication of diseases in the past, it no longer is appropriate for the devastating impact of MSK related disorders. This may be because of biomedicine's focus on a cure of a disease based upon the implicit belief of monocausality (single cause, single effect), linear causal relationships found amongst statistical frequencies and the lack of acknowledgement of context. In fact, the maintenance of biomedical causal beliefs in regards to pain has been found to be a barrier in the recovery of patients with low back pain (Bunzli et al. 2016). This may be because these patients are still looking for 'the' cure or the scan that will identify 'the' cause which will lead to the correct treatment and cure. This often feeds a self-fulfilling cycle of negative emotion, despondency, frustration and anger that also adds to the pain experience.

The mutual understanding and respect of the patient creates the cornerstone of positive change and the opportunity for restoring wellbeing. The patient's context, not the statistical average, guideline or diagnostic classification, is the key to improving the challenges that patients with MSK conditions have. As clinicians we are, after all, here to help people through the difficult challenges of life and, in my mind at least, there is no better purpose in life than to help others. I mentioned earlier that being a physical therapist is like being a detective, but it is really much more than that. Physical therapists are coaches, mentors, and care providers, but most importantly, they are relatable humans who can provide the structure and therapeutic environment for people to recover, heal and grow. Getting to the heart of the matter is more than statistical observations and analysis of average treatment effect sizes; it is the embodiment of human understanding, genuine curiousness, purposeful support and thoughtful guidance towards the emancipatory experience of agency and the fulfilment of our patient's goals. It is our patient's stories that ought to be honoured, respected and understood as the centrepiece for clinical decision making, with population data, clinical research, policy and guidelines used to support, rather than to dictate person centred care.

References and Further Readings

Babatunde F, MacDermid J, MacIntyre N (2017) Characteristics of therapeutic alliance in musculoskeletal physiotherapy and occupational therapy practice: a scoping review of the literature. BMC Health Serv Res 17:375

Barker K, Reid M, Minns Lowe J (2009) Divided by a common language? A qualitative study exploring the use of language by health professionals treating back pain. BMC Musculoskelet Disord 123:1–10

Bunzli S, McEvoy S, Dankaerts W, O'Sullivan P, O'Sullivan K (2016) Patient perspectives on participation in cognitive functional therapy for chronic low back pain. Phys Ther 96:1397–1407

Djulbegovic B, Guyatt GH (2017) Progress in evidence-based medicine: a quarter century on. Lancet 390:415–423

Holden G (1992) The relationship of self-efficacy appraisals to subsequent health related outcomes. Soc Work Health Care 16:53–93

Keedy NH, Keffala VJ, Altmaier EM, Chen JJ (2014) Health locus of control and self-efficacy predict back pain rehabilitation outcomes. Iowa Orthop J 34:158–165

Kinney M, Seider J, Floyd A, Beaty K, Coughlin K, Dyal M, Clewley D (2018) The impact of therapeutic alliance in physical therapy for chronic musculoskeletal pain: a systematic review of the literature. Physiother Theory Pract 28:1–13

Koes BW, van Tulder MW, Thomas S (2006) Diagnosis and treatment of low back pain. BMJ 332:1430–1434

Launer J (2018) Narrative based practice in health and social care: conversations inviting change. Routledge, London

Low M (2017) A novel clinical framework: the use of dispositions in clinical practice. A person centred approach. J Eval Clin Pract 23:1062–1070

Low M (2018) Managing complexity in musculoskeletal conditions: reflections from a physiotherapist. In Touch 164:22–28

NHS England. https://www.england.nhs.uk/ourwork/ltc-op-eolc/ltc-eolc/our-work-on-long-term-conditions/si-areas/musculoskeletal/, https://www.england.nhs.uk/ourwork/ltc-op-eolc/ltc-eolc/resources-for-long-term-conditions/

Øberg G, Normann B, Gallagher S (2015) Embodied-enactive clinical reasoning in physical therapy. Physiother Theory Pract 21:244–252

Open Access This chapter is licensed under the terms of the Creative Commons Attribution 4.0 International License (http://creativecommons.org/licenses/by/4.0/), which permits use, sharing, adaptation, distribution and reproduction in any medium or format, as long as you give appropriate credit to the original author(s) and the source, provide a link to the Creative Commons license and indicate if changes were made.

The images or other third party material in this chapter are included in the chapter's Creative Commons license, unless indicated otherwise in a credit line to the material. If material is not included in the chapter's Creative Commons license and your intended use is not permitted by statutory regulation or exceeds the permitted use, you will need to obtain permission directly from the copyright holder.

Chapter 9
Causality and Dispositionality in Medical Practice

Ivor Ralph Edwards

9.1 Some Background

I am a clinically qualified general consultant physician in general medicine and clinical pharmacology in the UK, with over 20 years of clinical experience. I have held professorial chairs in both subjects, in Zimbabwe and New Zealand respectively. In New Zealand I was also Director of the National Toxicology Group and had 10 years of experience in drug regulation and pharmacovigilance as Medical Assessor for the New Zealand Medicines Adverse Reaction Committee and the Medicines Assessment Advisory Committee. I also had similar responsibilities to the New Zealand Toxic Substances Board for the registration of other chemicals and for advice on chemical safety. I was appointed Chairman of the Advisory Committee to the World Health Organisation (WHO) Programme for International Drug Monitoring whilst in New Zealand. Following the above, for the last 25 years I was the founding Director of the Uppsala Monitoring Centre, which is responsible for the technical support to all WHO supported drug safety monitoring programmes worldwide. In that role, I was the first to develop data mining in pharmacovigilance and risk-benefit analysis. I have written over 350 scientific papers and, whilst I am now retired, I am still actively involved in teaching and projects. In all this work, the challenge of determining causality in individual patients has been paramount.

A very important issue I investigated many years ago were reports of blindness in 7 patients from around Germany, reported as possibly being due to omeprazole, a drug used to treat peptic ulcer. It was a very wide selling drug and the pharmaceutical company was particularly concerned: because of the very high sales, chance associations with blindness were possible, but it was odd that the reports were all from Germany, and besides, they were from intensive care units around Germany. The first issue that I found was that a trial use of the drug (unknown to both the

I. R. Edwards (✉)
Uppsala Monitoring Center for International Drug Monitoring, Uppsala, Sweden

pharma company and regulators) was being conducted on the prevention of peptic 'stress' ulcers in patients in intensive care. Then I found that about half the seven patients had taken methanol overdoses, a well-known cause for blindness. The other three were diagnosed as ischaemic optic neuropathy, which is a complication of life threatening trauma or other illness affecting the eye circulation by hypotension, which was the case for these other patients. Since literally millions of patients from around the world had used the drug without reporting blindness, and there was no known toxicological or pharmacological reason for blindness, I thought the other possible causes were most likely, in spite of the one unusual feature about these cases: they were receiving omeprazole intravenously and not by the far more common oral route. About 20 years further on I find that in huge longitudinal patient databases of over a million exposures it is suggested that blindness occurs in a few patients on omeprazole, who are predominantly female, over 60 and have hypertension: but this is another group in which blindness could be due to other causes.

Am I wrong to think that when I can see a more likely cause with a known mechanistic explanation – toxic methanol, prolonged and profound hypotension – that those are more likely the causes? And the damage from hypertension and other changes to eyesight in an aging population is at least as likely an explanation. But why especially women? There is clearly some uncertainty about my causal diagnoses here.

9.2 Considering Causality

Always we need the best information possible; we need to be transparent in our reasoning; we need to follow up and we must be open to new evidence. Clinical medicine uses multiple ways of accomplishing its ends, which are the diagnosis, treatment, amelioration or cure and, if possible, prevention of any dis-ease causing problems in an individual. At the heart of clinical medicine is an empathy with the patient in understanding what ails them in as deep and broad a way as possible. To do this, a practitioner uses, or a coordinated group of practitioners use science, art, learning and experience, and indeed whatever wisdom they can effectively bring to bear on a problem. A practicing clinician is not a scientist per se, but rather *uses* science.

Diagnosis is an essential first step. Some illnesses are easy to recognise, but even with those, there are competing possible diagnoses. Illnesses are shape-changing masquerades. Only a careful case history and examination of the patient will give us a useful picture of a range of clinical conditions for further consideration using observations and tests. A diagnostic assessment usually produces a list of possible causes to explain the patient's clinical signs and symptoms and their chronology, and that can be listed in order of probability in any given patient and in their particular context. The key challenge is what might be the cause of this patient's clinical state, but it is not only the various disease states (medically recognised clinical illnesses rather than their phenomenological consequences) that need to be considered, but also how that individual patient will react to a given dis-ease entity (dis-ease being the symptoms, signs, anxieties and any other personal consequences that are concerning them). A diagnosis must take into account not only context (environmental and familial) and the various disease attributes but also those proclivities of the patient in how they respond to dis-ease challenges.

A classic triad of questions is useful:

Can a disease ever cause this clinical state?
Has it done so in other humans?
Is it responsible for the clinical picture in this particular patient?

The essence of these questions is to understand the phenomenology of the disease and then to consider causality. Some would argue that consideration of the broader phenomenological aspects (their dis-ease) – the overall impact of disease – has little or no place in determining its cause. But this is to disregard the ways in which the patient's personal background and context may influence the ways in which they present the features of their dis-ease to the clinician, and in turn how the clinician interprets those features into different categories (physical, social, psychological, spiritual etc.) and then cares for and manages them. Variations in diseases themselves and their presentations in patients may alone lead to considerable misconceptions, for example about the severity and nature of pain, or indeed the description and localisation of any symptoms within their bodies.

Commonly, we *make* diagnoses. We look for the cause of an illness, or at least for an explanation of the physical signs and symptoms the patient presents. We are limited in interpreting those pieces of clinical information by what we have remembered about disease entities and the ways in which humans respond to them. We are, however, able to use many other information sources to aid our thinking, although it will take time, experience and imagination to find all the relevant material. We may have an easy task with a clearly recognisable pattern we commonly see in our daily experience, but if we find that there are dis-ease components in the pattern (due to individual dispositions) we do not recognise, this leads to confusions. Consider a patient whose family history is one of close members having had coronary artery disease. If that patient should have chest pain, she will naturally first think that her chest pain is due to heart disease and indeed may be more likely to describe her pain as having anginal qualities, perhaps ignoring the exact position of the pain in her chest. In turn, the patient's context is ever changing and needs to be considered while making a diagnosis and subsequently during their treatment. It is at this point that the dispositions of the disease should be considered as being modified by those dispositions of the patient.

Sometimes the patterns are complex and responding to their intricacy can be very demanding. Patients are not just carriers of disease entities, they have their own dispositions which react to disease differently and also to the same disease at different times and in other situations. Dispositionality, as a way of probing the phenomenon of a person's dis-ease seems to be a useful way of analysing a clinical situation. In respect to the example just given, we now know that women with coronary artery disease present a different spectrum of symptoms compared to men.

I and some colleagues were concerned by the roll out of WHO's '3 by 5' Programme in 2003, which aimed to provide AIDS treatment to 3 million people in Africa by 2005. It was a laudable and ambitious programme, but we were concerned about the little attention to monitoring possibilities for effectiveness and safety of the known toxic treatments. I attended a meeting in South Africa soon after the launch and there were several reports of deaths from lactic acidosis, a result of mitochondrial damage. *It was noted that those early South African cases were all women.* The drugs concerned were part of the 'highly active antiretroviral regime'. Stavudine and didanosine were the main suspected drugs that had been noted to cause this problem, which was rarely seen in the US and Europe, *where it was usually only symptomatic and reversible.* Now we know that around 80–90% of cases of lactic acidosis occur in women (in Africa) and that their mortality is as high as 50%. The early experience of these drugs in the western world was largely in men, and an early paper (Brinkman 2001) promoted the idea that there was small incidence with no serious clinical significance. We therefore knew of the potential problem from before the WHO roll out, but epidemiology in males suggested a small risk.

More recent reviews (e.g. Trang et al. 2015) explain more about the mechanisms. But there is more: it is obese African women with a high body mass index that are particularly at risk. Why? Also unknown is whether there is a possible genetic cause in African women. Are there other dispositional reasons why these patients are so badly affected? Are there contextual problems, such as poor availability of lactic acid screening for early symptomatic patients, that makes the mortality so high?

Suppose the clinician elicits a full case history (and this supposes no time constraint) and does a complete relevant physical examination. Each of the findings may be a disposition of a disease entity, or of the patient reacting to their situation, or may indeed be a general disposition of that patient. For example, does one patient always look pale and possibly anaemic, or perhaps another may have so called 'white coat hypertension' (raised blood pressure whenever it is measured in a hospital environment)? We may be faced with a patient that is garrulous and often chatters inconsequentially. I once had such a patient with a large open, varicose ulcer on the leg, and a mild fever. Her son accompanying her was a senior clinician himself, who apologised saying, "She's always chatty like this." I accepted this, but decided to admit her since she needed both analgesia and antibiotics: it was also late in the evening. The following day she had a very high fever, was very drowsy and had a falling blood pressure. This was a septicaemia with delirium and the knowledge of such a possibility should have alerted me to investigate more thoroughly whether infection was starting to cause a delirium. I was swayed by my colleague's reassurances with near disastrous consequences. She had a perilous passage through intensive care with a severe septic shock. Embarrassingly both mother and son were grateful to me.

In the past and now also in the present, many seek a single cause for an effect. It seems, however, much more useful to think of possible causes and to understand their mechanisms, and so consider a range of probabilities of causality based on the situation for a particular patient. A sore throat is likely to be infectious if a person has been enclosed in a crowded space during a winter epidemic but one would be more likely to consider a drug cause (e.g. the much rarer agranulocytosis) than if the sore throat occurred de novo while the patient was convalescing alone and already taking an antibiotic.

This clinical consideration of patients demands a broader view, deeper thinking and longer timeframes than most epidemiological studies or clinical trials allow in providing evidence useful in clinical practice. Clinical trials tend to focus on a simple connection between cause and effect – the likelihood of drug X being strongly associated with effect Y for a statistically significant proportion of those who take it, compared with controls. But a strong *association* is not proof of causality on its own, it remains a strong probability only. Nor is probability proof of a *cause*: it is essential to couple cause with effect by understanding the various ways in which attributes (or dispositions) of the disease, of the patient, and of the treatment and the context of the patient all interplay. Moreover, perfect, linear, causal relationships are rare in medicine, and the strength and variation of a disposition are as important as the fact of its mere presence or absence.

Causality will also become ever more important as we attempt to make the best use of genetic mechanisms behind the ways our bodies function. The new generation of gene therapies and other such 'personalized' treatments are more targeted to specific basic biomic functions in the body, which result in our dispositions and their strengths of expression. So, a proper, dispositional grasp of causality is a vital tool in helping healthcare professionals reach the best judgements, especially when time, resources and reliable information are in short supply. A keen understanding of all the factors underlying a clinical problem is the path to efficient use of resources, rather than the use of an overly simplistic but rigid 'guideline'; guidelines should reflect nuances of variation rather than simply dictating a pathway to a single algorithmic 'truth'. It is therefore better to consider a *causal explanation* of how a patient's symptoms and signs might appear as they do rather than to concentrate on the ideal of a single direct cause and effect. That is, to consider that the whole phenomenon of dis-ease includes the propensities of other dispositions to have an additive, augmenting effect as well as possible secondary effects.

Consider for instance a hypertensive patient, treated with beta-blockers, who dies from anaphylaxis after a bee sting or another allergen such as penicillin. The beta-blocker may well have contributed to the causal mechanism underlying the fatal event, by reducing the cardiovascular response to the severe hypotension caused by the acute allergic response. Similarly, the known sedative effects of a beta blocker may add to those of a benzodiazepine in a patient with a high blood pressure thought to be due to anxiety, with the result in a secondary effect of a fall with injuries. There would be a degree of speculation in such situations about what was in fact causative. Such speculations would probably not prove practically useful in the acute situation but they may give some chance of avoiding similar occurrences in the second case, and accumulating experiences of this kind might point towards a way to allow avoidance of important problems for patients in the future.

An elderly relative, taking warfarin anticoagulant after a series of minor strokes, developed heart failure and was treated with a frusemide, a diuretic commonly used to remove excess water from the body. He improved and his ankle swelling from the heart failure reduced. Some days later, he had a rash and painful ulcers on his lower legs and feet. The nursing home staff said they thought it was bed sores since he had been sitting and lying most of the time. I was sceptical because the ulcers were on the front of the feet and legs, not the right place at all. A dermatologist was consulted who suggested that the rash was a vasculitis, and I was able to suggest that frusemide was the cause, and the diuretic was changed. It was all too late; for he found the pain from the vasculitic ulcers so bad he needed morphine which made him sleepy and his breathing was also suppressed. He developed pneumonia and died peacefully from two adverse reactions to his treatments: the frusemide caused vasculitis, which caused pain, which caused morphine, which caused respiratory depression, which caused bronchopneumonia, which caused death. This kind of complex chain is quite common in medicine and illustrates why a causal explanation is valuable for understanding.

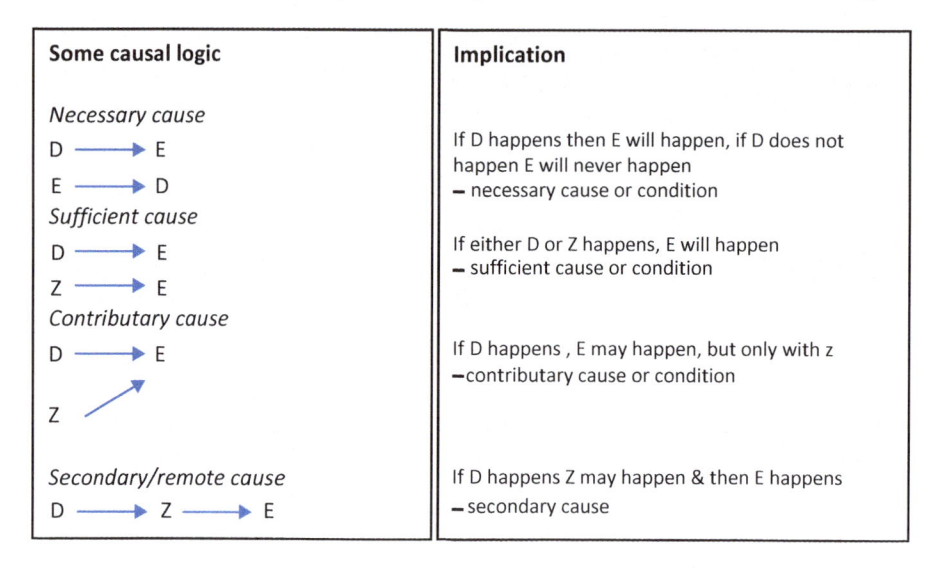

Some causal logic	Implication
Necessary cause D ⟶ E E ⟶ D	If D happens then E will happen, if D does not happen E will never happen − necessary cause or condition
Sufficient cause D ⟶ E Z ⟶ E	If either D or Z happens, E will happen − sufficient cause or condition
Contributary cause D ⟶ E Z ↗	If D happens, E may happen, but only with z −contributary cause or condition
Secondary/remote cause D ⟶ Z ⟶ E	If D happens Z may happen & then E happens − secondary cause

Most clinical healthcare practitioners would like to practice medicine with a detailed and empathetic diagnostic work-up including some of the considerations above, but also most of us know that time constraints do not allow for every patient to have a full assessment. In very many instances, such an approach is unnecessary and even counter-productive: emergency situations and the treatment of acute common diseases with generally good outcomes are examples. It is nevertheless wrong to consider one instance of contact with a patient in isolation as adequate. One meeting with a patient allows a preliminary assessment of immediately important dispositional factors. A patient meeting with her family medical practitioner for the first time for years may be asking for a symptomatic remedy for a persistent cold and cough and then mention a heavy period as an aside. The same patient may refuse an examination on the grounds that she is embarrassed because she is currently bleeding. Treating the symptoms of a cold without making an arrangement to properly pursue the vaginal bleeding would be a mistake indeed. It is very helpful to have continuity of care where a single clinician knows a family's background, and would be alert to behaviour that was unusual.

There are some instances where treatments are routinely commenced with a complete assessment of how an individual patient differs from the norm, followed by a carefully-balanced choice of medication bearing in mind the variability of the cause(s) and additionally keeping a close eye on how things turn out in the long term. The way of using antiretrovirals for HIV/AIDS is an obvious example. AIDS is a chronic disease which has various phases and states of ill-health (dis-ease), some induced by the changes, sometimes progression, of the disease process and sometimes due to treatments. Because the immune system is negatively affected, it also leaves the body open to various kinds of infections and different kinds of neoplasia. Patients change over time, as do their diseases and treatments, not only because of some disease processes but also because of aging. These patients and all people need careful monitoring in all aspects of *their* lives. Diseases and their management are all unique phenomena. But this sort of premium care is still unusual since it can only be delivered through healthcare systems with extensive personnel and technical resources linked to stable and well-organised health services.

For all patients, however, it seems best to consider *why* a particular clinical effect happens, *what* can be done about it, and *how* best to take action. This points to the need for a far more nuanced and holistic approach, which acknowledges that the way a treatment acts on an individual depends on their constituent dispositions towards different disease effects. It is about who those individuals are, where they are, their circumstances, history, what they eat and any other conditions or substances that affect their body systems. In aging societies, those older people are likely to suffer more illnesses, live with more chronic conditions and take more drugs to counteract them. This has profound implications for understanding causal relationships between dispositions. Successful management of a patient is best achieved by getting detailed information from, and on, an individual's dispositions, then matching it to the known characteristics (dispositions) of the treatment and then following up the patient to ensure that an optimal result has been gained.

Traditionally, medicines that demonstrate a high probability of achieving the desired outcome in controlled conditions are considered safe to market, prescribe and use. But in the real world, we know that even the best drugs typically only work as desired around 70% of the time. Variables driven by misdiagnosis, treatment variables such as dosage and compliance play out alongside environmental, genetic and individual factors to reduce the actual effectiveness, so therapeutics can only be improved by exploring risk and benefit probabilities and carefully monitoring outcomes, particularly of new treatments.

9.3 Diagnosis and Decisions

Causality in individual clinical decision situations is much too important to be left to chance or limited to the broad-brush norms defined by epidemiology. And we shouldn't let the cost or difficulty of pursuing the ideal deter us from doing what's right and good. Patients need and deserve nothing less.

Establishing a working diagnosis is the first major goal. The first focus of a clinician will be on the characteristic dispositions of the possible disease entities, arraying them in order of likelihood (differential diagnosis). This should also take account of the patient's context (environment) and the ways in which the human body tends to respond to the disease to produce symptoms. There are likely to be uncertainties due to gaps in information, or to variation in the strength or likelihood of features (dispositions) of the disease or of the dispositions of the patient such as response to pain or blood loss or immunity. What is the most likely causal link that explains the patient's symptoms and signs – qualitatively as well as quantitatively? What important data is missing and must necessarily be found before one can decide on a plan of action? What are the key dispositions we can use to follow the progress of the patient and the disease entity? For example, a microbiological identification and presence of bacteria in different body tissues or excreta usually enables us to decide on the dispositions of the likely causative organisms. We might, however, need to act without knowing the precise nature of the infection. But we can start treatment before the microbiological tests are done, then we must follow up the effect on the patient carefully, which includes measuring the temperature and heart rate of the patient and any other key dispositional responses ('vital signs') that may change, as well as measuring the success or otherwise of the patient's response to the infection. We also see that therapy and management have dispositions as well: the chemical structure, pharmacology and toxicology of any medicinal product have their dispositions, good and bad. They too interact with those of the patient and disease entity. Management may include other therapies than drugs such as chest physiotherapy to expel unwanted secretions in acute bronchitis or pneumonia.

A proper, dispositional grasp of causality is a tool to help clinicians reach the best judgements, especially when time, resources and reliable information are in short supply. Dispositional thinking is a dynamic way of sifting evidence about both disease and patient.

Many clinicians will think that talking about dispositions adds nothing to the way they already do their work, and indeed that dispositionality brings confusions. Many clinicians also feel that they have enough experience and intuition to pick up nuances of both patients and disease behaviour that are outside the norm – and many can. The stress, however, of work, of time pressure, of limited resources, can lead to mistakes being made. Thinking dispositionally can provide a way of double checking what we do and highlighting uncertainties that are inherent in diagnosis and management decisions in medicine.

Thinking about the dispositions of both patient and disease leads both to completeness and clarity in management. For example, the patient may be an aging alcoholic (dispositions to check might be liver and kidney function etc.) and the drug might be toxic at higher dose levels (so considering how it is metabolised and excreted might lead to a lower starting dose). Recognising these factors might also lead to the necessity of following up the patient after a given time to check on the patient's progress, perhaps with appropriate tests. Other less obvious considerations may follow from these more obvious ones, such as checking the patient's memory (any early dementia), their eating habits (if the drug should be taken with food) or their daily habits (if the drug is a diuretic they should be informed about the likely time of the diuresis, and plan to be near toilet facilities during that time).

9.4 Overview of Important Dispositional Insights in Clinical Care

1. Personal attitudes toward the patient can influence one's assessment and actions. This is an underused but key matter and a useful test of one own overall disposition in relation to the patient.
   - Do I like this patient? Do I find that the patient smells? Are they condescending and impolite? Or engaging and lively?
   - There are many more sophisticated possibilities and eliciting them will allow a critical view of factors that need to be considered and allowed for in the patient interaction.

2. What are the clinical dis-ease symptoms and signs in this patient?
   - Does the patient look ill or in pain? Are they afraid? Are they embarrassed? Impatient? Are they hesitant or in any way unclear in their responses?
   - These are dispositional features that must be taken into account in the interaction with the patient as well as being of diagnostic import.

3. Checking the clinical findings about diseases against prior clinical knowledge: what is key, what is missing, and what is unusual?
   - Assumptions dependant on scholarly descriptions of diseases and on experience can be limited in scope and misleading. A conscious check on the dispositions presented by the patient can help avoid premature assumptions.
     (i) As the patient's story unfolds and physical signs are elicited, the clinician will be alerted to a range of diagnostic possibilities to check against known features of diseases. Missing data must be considered carefully with each potential diagnosis. The variation in power of disease features needs to be considered against the array of responses possible given the dispositions of the patient. Consider an easily understood example from my own experiences of investigating pain. The *type* of pain is tricky, as it can be challenging to determine across cultures. In Zimbabwe what many would say is a 'stabbing pain' is described as 'pricking', thus raising questions about how severe a pain is. Even when the site of a pain is described, it may confuse a clinical appraisal because of the anatomy of the nervous system where a nerve branches, and disease affecting one branch is felt in an area innervated by another (referred pain). This may cause confusions in dental pathology, for example.
     (ii) The patterns of symptoms and signs elicited may overlap for a number of diseases. The immune responses to infections, to cancers and to other invasions by entities recognised as foreign to the human body can provide examples of this kind of confusing situation. Spontaneous abortions may be due to the mother's body recognising the fetus as a foreign invader, for example. Infections produce a change in the immune system that result in

normal body tissues being seen as foreign, in autoimmune disease (although this may be triggered in other ways, too).
(iii) Consideration of the interplay between various dispositions of the disease and its unwilling host, leads us to analyses of powers of diseases, their modes of expression and their actions. In turn this gives us ideas about useful laboratory tests and further monitoring of the patient's situation including the kinds of treatments that might be useful.

Considering dispositions also leads to considerations about what other evidence might be available to understand the disease and treatments and particularly in understanding cause and effect relationships.

4. Specific help can be obtained from the Bradford Hill proposals (Hill 1965) as well as DoTs (Aronson and Ferner 2003) and EIDOS guidelines (Ferner and Aronson 2010).

5. It is useful to think about dispositions in considering the value of evidence available from other sources, and particularly statistical associations. The choices of keywords in searches are usefully specified using dispositions, for example.

   – Decisions in managing patients such as choosing a medication are critically determined by choosing the correct medicinal product to be active against the disease, but not to do harm to the particular patient, who may be sensitive and consequently harmed by one product and not another.
   – The weighing of effectiveness against risks makes any such decision challenging. A dispositional approach to an individual's possible idiosyncrasies and much more dispositional information on the effectiveness - risk profiles of medicines as they are used in routine clinical practice, as opposed to controlled studies, is needed.

6. Interactions between dispositions is important where multiple diseases might or do coexist in patients, or when multiple therapies are in use.

   – Whilst commonly found in older patients, there is an increasing propensity for multiple disease/ treatment situations to develop as more medical disease situations are recognised.

7. Causal explanations, including possible interactions between dispositions, are of considerable value clinically even if speculative: the speculations are enhanced with other similar occurrences and by prior scientific or clinical evidence of the powers of those dispositions. Any instances of interactions and unusual outcomes should be reported in detail and be made available for others to share.

8. Many individual and rare medical situations fall outside the norms usually considered in controlled clinical trials and even controlled observational studies. Thinking about variations in the disease and patient dispositions possible through theoretical and practical knowledge of mechanisms is important.

- Rare disease presentations and unusual adverse reactions may not occur often in relation to an individual disease or medication, but the totality of such examples over the whole of healthcare is huge. Adverse drug reactions were the 5th most frequent cause of death in the US and similar findings have been seen in several other countries.
9. As we learn more about genomics and biomics, the links to the incidence and powers of dispositions will be very important in diagnosis and therapeutic decisions.
   - Linking statements 6 and 7 emphasises the value of individual case reports with full descriptive detail to allow medicine to progress.
10. *Causal explanation* needs to be more widely practised to allow us to understand better how the propensities of different dispositions of disease, patient and treatment interact for better or worse outcomes.
    - Rejecting the inclusion of subjects with identifiable potentially confounding propensities is a two-edged sword: it allows for clarity in identifying a possible statistical association but removes the potential for multivariate analysis.
    - Statistical association and particularly non-association should always be tempered by what we know about the possible mechanisms by which a proposed cause could produce an effect, plus any other dispositional evidence demonstrating why a particular instance of cause and effect might be rare or unique.

9.5 Conclusion

Humans share many attributes but there are many examples of unusual ('orphan') diseases as well as rare adverse reactions to therapies. These situations need dispositional thinking and not only epidemiological, normative approaches. Having had the privilege of being a clinician as well as working in a scientific setting experimenting with drugs and chemicals, I can conclude that medicine is not a science but the application of described knowledge or knowledge acquired by experience. Medicine is its own discipline in which the essential skills are:

1. For the clinician to match the dispositions of the patient in front of them with all their acquired knowledge of others who have similar dispositions, to find the closest binary match and to understand the probable constitutions of their patient.
2. The clinician must understand the dispositional weaknesses of their particular disease(s) and so, choose a treatment that has the maximum benefit for the patient with the least harm.
3. All of this must be undertaken with a specific aim of treatment that is in agreement with the patient and takes account of the patient's social context using an empathetic and holistic (phenomenological) approach.

4. Clinicians also have responsibility to ensure that all their reasoning from diagnosis and clinical management works in real-life clinical practice. They must pass on to others their knowledge, particularly when outcomes are unexpected.
5. The responsible clinician should also try to identify *why* the unexpected outcome occurred, so adding to global knowledge.

It is clear that the above six points are most difficult to attain, but the vision should remain.

References and Further Readings

Aronson JK, Ferner RE (2003) Joining the DoTS: new approach to classifying adverse drug reactions. BMJ 327:1222–1225

Brinkman K (2001) Management of hyperlactatemia: no need for routine lactate measurements. AIDS 15:795–797

Ferner RE, Aronson JK (2010) EIDOS: a mechanistic classification of adverse drug effects. Drug Saf 33:15–23

Hill AB (1965) The environment and disease: association or causation? Proc R Soc Med 58:295–300

Trang AQ, Xu LHR, Moea OW (2015) Drug-induced metabolic acidosis. F1000 Faculty Rev. https://doi.org/10.12688/f1000research.7006.1. PMCID: PMC4754009, PMID: 26918138

Open Access This chapter is licensed under the terms of the Creative Commons Attribution 4.0 International License (http://creativecommons.org/licenses/by/4.0/), which permits use, sharing, adaptation, distribution and reproduction in any medium or format, as long as you give appropriate credit to the original author(s) and the source, provide a link to the Creative Commons license and indicate if changes were made.

The images or other third party material in this chapter are included in the chapter's Creative Commons license, unless indicated otherwise in a credit line to the material. If material is not included in the chapter's Creative Commons license and your intended use is not permitted by statutory regulation or exceeds the permitted use, you will need to obtain permission directly from the copyright holder.

Chapter 10
Lessons on Causality from Clinical Encounters with Severely Obese Patients

Kai Brynjar Hagen

10.1 Introduction

This contribution is about how I came to value and implement the search for a genuinely causal diagnosis in a specific group of patients, the severely obese. When I started to work as a senior consultant at the Regional Centre for Morbid Obesity at Bodø Hospital (RSSO), my background in General Practice (GP) was unusual for the position. The RSSO is a specialist hospital centre assigned to endocrinology within internal medicine, and closely cooperating with surgery. I soon became distressed in my professional role. It seemed to me that the healthcare system in which I was supposed to play my part provided only rather shallow symptomatic diagnosis and therapy. I felt like (one of) "[…] those who are troubled by the disparity between the formal biomedical diagnoses we learned so proudly and the actual human problems that patients bring to us in our offices" (Felitti 2003: 84).

The reason for my distress was that I had decided not to mimic the specialty of endocrinology, but to stick to my GP-approach, conceptualised as whole person care (WPC). This concept has recently been reviewed and described as follows: "a multidimensional, integrated approach; the importance of the therapeutic relationship; acknowledging doctors' humanity; recognising patients' individual personhood; viewing health as more than absence of disease; and employing a range of treatment modalities" (Thomas et al. 2018: 1). The study suggests that "GPs understand WPC to be an approach that considers multiple dimensions of the patient and their context, including biological, psychological, social and possibly spiritual and ecological factors, and addresses these in an integrated fashion that keeps sight of the whole" (Thomas et al. 2018: 8).

I encountered several discrepancies between my whole person approach and the specialist hospital setting. First, the specialist routine was to look at one part of the

---

K. B. Hagen (✉)
Regional Centre for Morbid Obesity, Bodø, Norway
e-mail: kai.brynjar.hagen@bodo.kommune.no

patient at a time, not the whole person. Second, the obesity centre's approach relied heavily on traditional biomedical practice, reflected in authoritative guidelines for clinical endocrinology and focusing on the somatic comorbidities of obesity (cf. Garvey et al. 2016). There was thus little room for attention to life story and psychosocial conditions. Third, I found that throughout the health system, negative attitudes toward obese people were widespread, assigning them low status and low priority, just as American psychiatrist Hilde Bruch noted almost 80 years ago: "Overeating is looked upon as a moral weakness and self-indulgence. Even physicians may express a sarcastic attitude" (Bruch 1948: 84). Stigmatisation of obese people is paramount, considerably affecting the quality of their care (Phelan et al. 2015; Williams and Annandale 2018).

Were negative attitudes the reason why obese patients with eating disorders were declined from getting help at the Regional Centre for Eating Disorders? Obesity was not listed as a diagnosis in line with anorexia nervosa, and would not be funded as much (less income for the hospital). Consequently, a large group of patients, the obese ones – many of them with a type of background similar to that of their thin counterparts – were simply excluded from this part of the psychiatric department. Referral to trauma treatment was also in many cases declined. The reason that was given was that the patients did not have symptoms of severe psychiatric disease. No, they were not mentally ill, but they had an unbearable feeling of emotional pain inflicted on them. In the manuals of psychiatry diagnosis, there seemed to be no slot for obese patients with emotional pain.

My impression was that healthcare professionals assumed the obese patients to "carry some heavy luggage". These assumptions, however, were neither investigated in depth nor given relevance in the person's records. The healthcare system seemed ignorant of the significance of a causal diagnosis in cases of severe obesity and indifferent towards scientific evidence indicating a potential for improvement when considering the impact of lifetime adversity on health.

Luckily, at RSSO, I am part of a competent interdisciplinary team, including specialist nurses, colleague doctors and a clinical nutritionist as well as surgeons, who have been open to my suggestions to increase the focus on psychosocial factors and adverse life events. In this collaborative environment we have increasingly turned away from a purely biomedical perspective and toward a whole person view on the patients. Discussions within this team have since provided valuable new perspectives on diagnosis, therapy and follow up. We soon realised that the framework for the clinical encounter was one of the things that we could quite easily change. I will here describe the approach with which we are now trying to get closer to causal diagnosis for patients with severe obesity at our centre RSSO, and briefly present three cases from my work there.

10.2 A Framework for the Clinical Encounter

If one is truly to succeed in leading a person to a specific place, one must first and foremost take care to find him where he is and begin there. This is the secret in the entire art of helping. Søren Kierkegaard, 1880
(Kierkegaard et al. 1998)

10.2.1 The Person in the Role of the Patient – What Are the Goals of Healthcare?

The role of the patient implies a specific status, framed by legislation, informed by rules and norms, and endowed with rights and duties for the healthcare system as well as for the patient. Every person has unique reasons for entering the role of the patient – which is not always an easy decision. Typically, the outcome of medical intervention depends on the clarification of the reasons for seeking help.

In general practice, learning to identify the patient's goals is mandatory. Consultants seeing patients with obesity in a hospital setting face a similar challenge. Relevant information for understanding the patient's goals might be found in the referral letter. In addition, patients are asked to fill in a form including questions concerning their goals, and how RSSO might help. Reviewing these goals is important at all stages of the further process: at the preparatory seminar, at the clinical encounter and during the follow-up period. Perceptions might change – patients' as well as doctors' – allowing for new perspectives, knowledge and emotional maturing.

Examples of goals may be:

- stopping further weight gain
- losing weight, sometimes specified in kilos or percentage, in some cases aiming at a "normal" Body Mass Index (BMI)
- reducing comorbidities like sleep apnoea, diabetes, hypertension or joint pain
- facilitating improved social participation and avoiding stigmatisation
- improving physical function like hiking in the mountains
- meeting the requirements for a specific job
- enhancing the general quality of life
- being able to spend time with grandchildren.

The patient's expectation of help from RSSO, initially simply being bariatric surgery, may become more differentiated during the course of the consultation process. Analogously, the clinical team's understanding of a patient might improve, enabling them to offer more well-targeted help.

10.2.2 A Group Seminar Before the Clinical Encounter: Setting the Stage

Information about the patient is initially collected from the referral letter and the form. Further preparations for the clinical encounter are made a couple of weeks in advance at a group seminar for 12 patients. At this seminar, members of the clinical team inform about the medical examination, treatment options, the diagnostic approach and therapeutic strategies. They also provide an update on the available knowledge, stress the necessity of avoiding conflicting objectives, and give the patients the opportunity to discuss with them. During the seminar, the patients and the team members become more familiar with each other's attitudes and priorities.

The information the team conveys at the preparatory meeting includes the following points:

- Everyday physical activity on a sustainable level is recommendable, but physical training normally cannot compensate for too high energy intake.
- Nutrition issues are discussed, especially the discrepancy between cognitive knowledge and actual practice, which means that emotions often play a major part in determining eating behaviour. A better mental condition may make it easier to practice nutritional knowledge.
- The significance of bio-rhythms is outlined, especially the fact that regular circadian rhythm and meal rhythm is physiologically preferable.
- The emerging scientific understanding of the microbiota functions is mentioned, and also the epigenetic principles.
- We acknowledge that people with obesity might be exposed to different sorts of bullying and violations of their integrity, which makes active self-defence strategies important.
- Harmful stress is probably one of the most important causal factors for becoming obese, linked to a variety of adverse life events. Adverse events can lead to mental and physiological overactivation and painful feelings, which to some extent can be relieved by overeating, called emotional eating. Understanding these mechanisms should replace moralisation. I give an example from my own experience of stressful situations during night duties: chocolate with marzipan makes me more relaxed. A brief stress-relief very much needed at the time, this effect would not be achieved with a green salad. A stressful situation for a couple of hours is one thing, chronic never-ending stress is something quite different—perhaps I would have to 'take' chocolate continuously for relief? That might be one possible mechanism for the onset of overweight. A number of other possible coping mechanisms are known, for instance drug abuse, self harming and anorexia. In my opinion, the emotional pain caused by traumatic stress is not a mental illness, rather a normal reaction to an abnormal strain.

We present the patients with our goal: the dialogue between patient and doctor is successful if it brings an insight that neither of them could have achieved alone. The dialogue aims at a deeper understanding of the patient's life history and the current situation to enhance the validity of the diagnosis, and thus pave the way for effective therapy. Patients are encouraged to be prepared for a rather thorough and comprehensive review of their life history, but of course only as far as it feels natural; no one will be subjected to pressure. We will ask certain questions, such as, for example: Which life events could have contributed to the onset of undesirable weight gain? How is the current situation maintained? In some cases, one single identifiable cause leads to a person becoming overweight; other cases are unclear and complicated. We underline the importance of a whole person view, the uniqueness of every individual and the complexity of causal factors, rendering comparisons between patients invalid. We explain that the weight graph will be plotted on a timeline to look for possible correlations with life events.

10.2.3 The Consultant's Understanding in Advance of the Clinical Encounter

In advance of the clinical encounter, I take a few minutes to review the available information about the patient, filling data in the template for the medical record. This template includes not only biomedical data, but also an extended part for the life story, especially psychosocial conditions, as well as data for the weight graph. Sources of information are the referral letter, the patient's preceding, written answers in a form, and results from blood tests or other supplementary tests. Sometimes I also make notes concerning my impressions during the group seminar.

I reflect upon what the patient's everyday life may be like, where he or she actually is. In a meta perspective, as a kind of self observation, I try to be conscious of my own position, my background, experience, personality, current situation and communicative ability, since all this will influence the dialogue, the dynamic interaction, and consequently also the picture of the patient's life. Hypothetically, another doctor in my place would probably have come to slightly different conclusions. In some cases patients came to me for follow-up, and it became quite clear that the description of the patient's life story did look quite different in my view compared to the written record from the original examination. An example of this might be the patient who, after bariatric surgery some years ago and initial weight loss, came to me as he now had the same weight as the maximum weight pre-surgery. In my view, this was because his traumatic life story had not been addressed, understood and taken into account. As healthcare professionals, we all have different colored glasses through which we see the patient. These aspects of clinical communication have not been part of my training from medical school, but evolved through the years of clinical practice, and heavily inspired by dispositionalism. Young doctors have had training in patient-doctor communication, but it seems to me that the aspect of being part of the causal profile of the patient has not been addressed.

10.2.4 The Clinical Encounter

At the start of the clinical encounter, I ask whether the patient feels comfortable, which is a way to pay respect to her or his current situation, and also to find a good starting point for the dialogue. I regularly ask: "Did you sleep well last night? How was your journey?" In some aircrafts on regional flights, severely obese people are denied access because the seats are too narrow. In addition, the winter in Northern Norway can make any journey challenging. I also used to ask: "Does it feel stressful for you to come here?" Often, the patients express mixed feelings, being happy for being admitted to the clinic, but feeling uncomfortable about addressing their obesity. The patient may recently have been ill or injured, or is worried about children back home, among other things.

I go on to inform the patient about my way of working: that I have a template with the topical themes, that I will type part of what is said as we go along so that correct expression can be preserved, that we will review blood analysis results and medication, that we will carry out a brief somatic examination, and that together, we will assess further follow-up options.

I ask the patient about allergies, surgery, pregnancies/births, diseases, known risk factors, heredity, social situation (family, education, work, social participation), natural functions (circadian rhythm, sleep duration/quality, urination, stools, menstruation), physical activity, dental status (does he or she suffer from odontofobia?), use of tobacco and alcohol. Sexual function is not routinely addressed, unless the patient takes the initiative.

10.2.5 As a Child, Did You Feel Safe at Home?

The main focus of the clinical encounter, however, is the life story. I usually ask the patient: "As a child, did you feel safe at home?" This is one of the themes that can open the discussion for stories of violence, drug abuse, parental mental illness, incest, traumatic loss of a person close to them, and so on. I became aware that there are many forms of violence. As Norwegian writer Yngve Hammerlin – a man with severe childhood violence experiences – has pointed out, violence is often not adequately understood by doctors. From the point of view of those who have experienced violence, the professional's response can thus signal desperate simplification and reductionism. Hammerlin suggests that the attitudes of healthcare professionals need to reflect a deeper understanding of the many types and consequences of violence (Hammerlin 2014). The wounds of an unsafe childhood are also described by Swedish authors, Josefsson and Linge (2011).

I explicitly ask about social participation with children of the same age, as well as physical activities, well-being at school and how school subjects were managed. At this point, there are often heart-breaking stories about bullying during many years at school, and of social isolation. From early childhood, via pre-school, adolescence, elementary school, high school, and into young adulthood, many facets form a picture of the person's development, of adverse and supportive events, of weaknesses and strengths, all in a whole person view. Relations to family members, partners or spouse, as well as certain social relations, are important determinants for life quality. Some patients find themselves in a marriage with a psychopathic person, only, after divorce, to be persecuted by the same person. Also, predators abusing children sexually are very seldom brought to justice, based on what patients tell me. Some report that their childhood predator still lives in the local community, and there is always a risk of accidentally meeting him, which provokes stress and fear.

We try to compare variations of the weight graph with changes in the psychosocial situation. Sometimes there seems to be a causal relation between adverse events, or stressful periods, and weight gain. Other forms of emotional regulation like self-harming, drug abuse or anorexia life style can have occurred in certain periods of life. A minority of the patients have had psychiatric therapy, which, however, rarely seems to have addressed adverse life events. Psychiatric healthcare is comprehensive and includes a large variety of diagnostic and therapeutic methods. It seems to me that it has been a widespread perception that trauma in the patient's history should not be mentioned, due to an estimated risk of re-traumatising, whereas other institutions practice methods with basic trauma understanding. One example of the latter is the Viken Senter in Northern Norway (www.vikensenter.no).

During the clinical encounter it sometimes feels like the dialogue exists as an independent unit, giving insight neither patient nor doctor could have acquired on their own, based on communication with mutual respect and engagement, and equally shared contributions.

Before ending the consultation, test results are carefully explained, medication is considered, somatic screening is done, we summarise and make a plan for further follow-up.

10.2.6 The Consultant's and Patient's Understanding After the Clinical Encounter

The patient's nutrition is thoroughly evaluated by the clinical nutritionist or a specialist nurse, either before or after the clinical encounter. In case bariatric surgery is an aim or an option, the surgeon will also see the patient. Later on, when the reports on nutrition and surgery assessment are complete, I summarise the conclusions in the record which is sent to the local hospital and the patient's GP, electronically available for the patient, and sometimes also sent as a referral to a rehabilitation centre or a psychiatric ward where basic understanding of adverse life events is practiced. The clinical team might discuss the case.

After 2 days in the clinic, patients have a normal follow-up period of 6 months during which they are supposed to have a phone consultation with one of the specialist nurses or the clinical nutritionist every 2 weeks, with the goal of modifying lifestyle to lose weight, often discussing psychosocial conditions. In some cases, additional consultations at the clinic are necessary. Further treatment can be more conservative or supported by bariatric surgery. The post-surgery programme has been extended to 5 years.

A locally based coordinated team can be beneficial during a comprehensive follow-up, involving the patient, the GP, and other professionals according to the needs of the individual patient. If needed, RSSO can take part in a meeting with the local coordinating team by phone or by video/Skype.

The post-encounter period gives a deeper understanding of the individual patient as well as the causal factors involved. Symptomatic diagnosis and therapy might be useful for a limited period of time, but our ambition is always the causal diagnosis and therapy.

In order to demonstrate how this intention might be achieved, three vignettes will be presented. The clinical encounter can be enriched and provide new insights in light of biographical information focusing on relational and social issues during the actual patient's lifetime from childhood through adolescence to the present state. The three anonymised persons whose stories unfold in the following have consented to the publication of their accounts in this book.

10.3 Case Stories

10.3.1 Olav Olsen, a Severely Obese Man

When enrolled and examined at the RSSO, Olav Olsen, 45 years old, weighs 123 kg, his Body Mass Index (BMI) is 40.6 kg/m^2 and the waist circumference is 132 cm. He relates that his maximum weight has been 126 kg, and that various attempts to reduce weight have always been succeeded by weight gain.

Olav has been diagnosed with gastrointestinal reflux, oesophagitis, high blood pressure, sleep apnoea, type II diabetes (insulin-regulated), bronchial asthma since childhood, Mb. Bechterew since age 21, osteopenia, allergy and eczema. Additionally, liver enzymes and blood glucose measures are above the upper norm, and he suffers from generalised muscular pain. Cognitive psychotherapy has recently ameliorated his previously incapacitating social anxiety. He uses on regular basis a wide range of medications. With regard to possible heredity, type II diabetes and coronary heart disease (father) in addition to asthma and hypertension (both parents) should be mentioned.

Olav is single, has received a disability pension since age 31 and lives with his parents in a rural area. Meals are typically prepared by his mother and characterised by being irregular and high in carbohydrates. He himself describes his eating habits as "always eating too much and never having a feeling of satiety".

Olav relates that his childhood home was a safe place until he was 9 years old. From then on and until age 15 years, he was sexually abused by some uncles and cousins. However, he started being overweight already as a pre-schooler, and physical activities with his peers gradually decreased until he was in 6th grade. He was not exposed to bullying at school and learned the school subjects reasonably well until the age of 12, when he became anxious and concentration problems arose. 14 years old, he was so afraid of speaking up in class that he often hid in the lavatory to avoid such exposures. At 16 he had become clearly obese (91 kg). He did not enter secondary school due to significant anxiety.

When 24 years old, Olav's GP referred him to an outpatient psychiatric clinic where he received considerable help, and by age 27 he felt ready to report his sexual abusers to the police. Six of his uncles and cousins were taken into custody. The legal process lasted for 4 years, but without a criminal charge due to insufficient evidence. Olav received, however, criminal injuries compensation. Olav's case was reported in all media, which caused him considerable distress. He rapidly gained weight during these years, and at age 28 his weight was 103 kg.

In accordance with this highly informative and detailed first encounter, the following measures concerning Olav's medical demands and further treatment were proposed: an interdisciplinary follow-up including psychosocial support, improved co-morbidity disease management, assistance for lifestyle changes via frequent telephone conversations and, having implemented lifestyle changes leading to 10% weight loss, possibly bariatric surgery.

10.3.2 Alma Almas, a Severely Obese Woman

Alma is 26 years old when she comes to the clinical encounter. Her weight is 127,5 kg, her BMI 41.6 kg/m^2, and her waist circumference 129 cm. Vitamin D deficiency is found by the blood tests. Alma has been diagnosed with polycystic ovary syndrome, a condition resulting in multifocal pains and bleeding related to the menstrual cycle. She reports that her father is obese as well, and that he has been diagnosed with diabetes.

Alma drinks only minimal amounts of alcoholic beverages and doesn't smoke but uses snuff (nicotine). She has lived with her boyfriend for the last 3 years but has no children. Her level of daily activity is high: housework and dog-walking for 45 min twice a day and exercising 75 min three times a week at a fitness centre. She has worked as an assistant caretaker for mentally impaired patients since she was 21 years old. Except for the previous year, she has worked at night, resulting in frequent change of her circadian rhythm, causing stress and insomnia, leading to the use of sleeping pills.

Alma reports a lack of care during childhood. Her mother had been raped as a child and has been suffering from fibromyalgia as an adult. Her 2 years younger sister had craved much of her parent's attention due to sleeping problems until age three, and as a result of this they were often exhausted. Alma's family, living in a rural area, had poor economic resources, implying, among other things, impaired nutrition for all family members. Her father, working in a grocery store, would bring home expired food for free. Sweets were locked up.

The onset of Alma being overweight came when she was 8 years old. Although being one of only five pupils in her school class, her dyslexia was undiagnosed until she was 11 years old. As she had exercise-induced asthma (also undiagnosed for a long time), her ability to participate in physical activities was limited. Her social life as a child was poor, partly because of her habit to withdraw from the others and preference for being by herself.

Alma recalls that she at age 13, when spending her holidays at her paternal grandmother's, experienced a severe food restriction, leaving her constantly hungry for 4 weeks. This was a relational trauma because of the pressure she was exposed to, as well as a sort of somatic trauma related to starving. Later she learned that her grandmother had previously been hospitalised due to anorexia. When reporting this experience, Alma is visibly emotional.

She is almost amnestic for the time from age 13 to 18 years of age. She assumes that she must have suffered from severe depression. During these years she consumed large quantities of sweets and gained weight continuously. When 18 years old, she left home to attend a college in a larger town while also working part-time in a grocery shop. One year later, she experienced severe symptoms of burnout and could neither work nor study for 1 year. Frequent psychotherapeutic consultations over 6 months were helpful.

At 19 she had a break-up with her then boyfriend. At 22 she was assaulted on her way home from work by one of the mentally impaired patients; he had a knife, but she managed to escape.

Alma was informed about different treatment options, and has had a quite normal follow-up at the RSSO, consisting of mainly telephone consultations with the specialist nurse and no specific psychiatric therapy.

10.3.3 *Ebba Eskil, a Severely Obese and Depressed Woman*

At the time of medical examination at the RSSO, Ebba is 46 years old. Her weight is 106 kg, her BMI is 39 kg/m^2, her waist circumference is 131 cm. She has been diagnosed with type II diabetes at age 27, and her blood glucose is not well regulated. In addition, tests indicate vitamin D deficiency. She suffers from sleep apnoea and was treated with a positive airway pressure ventilator; continuous positive airway pressure (the positive air pressure made by a fan reaches the throat and keeps the airways open through the night). Furthermore, she has chronic lower back pain, neuropathic pain in both legs, oesophagitis and frozen shoulder. She was diagnosed with depression at age 42 and takes antidepressants. Her diet is high in carbohydrate and fat, with emotional eating and sometimes losing control. Previously a cleaner, she had been unable to work the last 4 years due to myalgia, diabetes and obsessive-compulsive disorder (OCD), and is receiving disability benefits.

She had surgery for an extrauterine pregnancy at age 25, gave birth to her first child at age 27 (vaginally), and to her second and third child by means of Caesarean sections at age 29 and 35 years. She lives together with her husband and three adolescent children.

As a child, she always felt unsafe at home. Her parents often had terrifying conflicts. Her father once cut a picture on the wall to pieces with a knife. She had one older and one younger sister; the children normally had to find food without help from either parent. Her relationship with her mother was especially difficult. The mother worked as a cleaner and forced her daughter to assist her at work, thereby isolating her socially from girls her own age. Her mother was addicted to gambling, frequently losing money. Her parents did not contribute to activities for children and parents at school. In addition, her dyslexia made for a difficult time at school. In the end, she had to take responsibility for her parents, being deprived of her own childhood.

Until she was 15–16 years of age, her body weight was normal. At 16, she moved out from home into a dorm. She rapidly gained weight, reaching a maximum of 126 kg. Due to increasing OCD symptoms she was referred to an outpatient psychiatric department.

A plan for care and follow-up was made, with focus on mental health support, improving nutrition and diabetes management, eventually bariatric surgery.

10.4 Where Do We Go from Here?

Her hair reminds me of a warm safe place
  Where as a child I'd hide
  And pray for the thunder
  And the rain
  To quietly pass me by
  Lyrics from "Sweet Child o' Mine" (Guns N' Roses)

For me, the clinical encounters have been a most valuable source of knowledge. In the dialogues with the patients, new insights have been opened concerning the causal factors for obesity. It has been a truly educational journey into the unique history of each individual.

As a child, did you feel safe? From January 2013 to July 2019, this question has been answered in 755 clinical encounters. The answers have revealed crucial adverse life events, each one unique, outrageous and challenging. The keywords are trauma, loss of a related person, sexual assault, deprivation of care and safety, exposure to violence and bullying. It seems especially harmful to be treated badly by one's mother. One patient told me that her mother said she wanted just three children, but this daughter was her fourth child, and she was treated accordingly.

I have come to reflect on some uncomfortable questions. How many people have been exposed to severe trauma or other adverse life events in childhood? In how many cases has trauma been succeeded by the onset of overweight, then bullying, social withdrawal, increasing weight and emerging comorbidities? How many other sequelae of trauma are prevalent?

Beyond the individuals, the clinical encounters have also provoked fundamental questions concerning health, disease and healthcare. How much injustice has been committed? How many people have known but not interfered? How many obese people have been economically exploited by the providers of the countless, aggressively promoted slimming products, symptomatic treatments and dubious surgery? What is the total amount of damage to the individual and to society? Are the national healthcare systems meeting obese persons in an adequate way?

Where do we go from here? In the next sections, I briefly outline the scientific evidence and make some suggestions for changes in clinical practice.

10.4.1 "What the Hell Is Going on Here?"

Yes, 'n' how many times can a man turn his head
  Pretending he just doesn't see?
  Lyrics from 'Blowin' in the Wind' (Bob Dylan)

Scientific evidence has at least since 1940 pointed to the importance of life stories for health and deepened our understanding of this causal relationship – although apparently without much impact on healthcare systems or clinical practice yet. Here are some episodes from medical science to illustrate the evidence.

In 1940, psychiatrist Hilde Bruch quoted Lichtwitz (1923) as observing rapidly developing obesity in women who had been under severe mental stress during and after the 1914–1918 war. However, it seems that this observation was not taken into the internal medical discussions on causality in the 1920's, when quite a lot of medical research was done on the causal factors of obesity. Life history was ignored, while biochemical and pathophysiological evidence known at the time seem to be thoroughly investigated in search for causality. Focus was on metabolic, endocrinologic and biochemical conditions. A salient conclusion was that obesity is not possible without malfunction of the central metabolic regulation (Bernhardt 1929).[1]

On the basis of her own work, Bruch described a home environment that did not offer adequate emotional security, and where food had gained an exaggerated importance; charged with a high emotional value, it represented love, security and satisfaction. The child may opt for the pleasures of food if it does not get the pleasure of love from its parents (Bruch and Touraine 1940). In 1949, psychiatrist

---

[1] This might be the roots of endocrinology being the branch of medicine responsible for obesity during the last century, as established in guidelines. The diagnosis of severe obesity in a traditional biomedical sense does not focus on life story, psychosocial conditions or adverse life events; however, somatic comorbidities are highlighted. An example of this is the authoritative American Association of Clinical Endocrinologists and American College of Endocrinology Comprehensive Clinical Practice Guidelines for Medical care of Patients with Obesity (Garvey et al. 2016).

H. J. Shorvon and GP John S. Richardson at St Thomas' Hospital in Cambridge, UK, described some important clinical observations:

> The investigation first arose when it was observed that some obese patients who failed to respond satisfactorily to the usual methods of treatment dated their obesity and also some psychoneurotic symptoms to a specific incident that might well have inflicted psychological trauma. We found that too little attention had been paid to the emotional factors in the first instance, and the patients had been treated on an organic basis for long periods. We do not claim that the treatment pursued with this group is primarily aimed at reducing weight, but we have used it to alleviate the associated mental distress and have found that patients often show a parallel reduction of weight and become more amenable to accepted methods of treatment of obesity. (Shorvon and Richardson 1949: 951)

In the 1980s, Vincent J. Felitti, an internist who founded the Department of Preventive Medicine at Kaiser Permanente in San Diego, made similar observations. He witnessed a paradoxical tendency for patients who had successfully lost weight in his ward to not participate in the follow-ups. He ended up asking himself: "What the hell is going on here?" (Kirkengen 2019). Felitti decided to ask these patients about their background, and he found that each of them had experienced considerable adversity (ibid.). In the "Adverse Childhood Experiences Study" (ACE-Study) Felitti and his colleagues later found a dose-response relationship between childhood trauma and obesity, as well as a range of other diseases (Felitti et al. 1998). More recently, a relationship in line with the ACE-Study was found in a Norwegian setting (Tomasdottir et al. 2015). A significant relationship between exposure to violence in adolescence and being overweight has also been demonstrated (Stensland et al. 2015). The relationship has been confirmed in recent meta-analyses (Danese and Tan 2014; Wang et al. 2015; Hemmingsson et al. 2014). A chronological relationship between adverse life events and the onset of obesity has been shown in a study by Lynch et al. (2018).

It is fairly well-established that emotional eating can be caused by trauma, depression and post-traumatic stress disorder (PTSD) (Talbot et al. 2013). The observations of the role of mental stress in Litchwitz (1923), and Bernhardt's (1929) focus on changes in brain physiology, both point to modern scientific theories of allostatic overload as a possible mechanism for the connection between trauma and obesity. Allostasis is "the process by which a state of internal, physiological equilibrium is maintained by an organism in response to actual or perceived environmental and psychological stressors."[2] Allostatic load involves psychological, neurological, endocrinological and immunological processes to cope with mental or physical challenges. Overload might result in dysregulation, and possibly disease (McEwen 2012).

Back at the RSSO, I asked myself why this scientific evidence has not been implemented in clinical practice. Why has accumulated knowledge from seven or eight decades not yet been acknowledged in healthcare? Even the modern, authoritative medical resource UpToDate does not mention very much about psychological

---

[2] Merriam-Webster online dictionary, entry on "allostasis". https://www.merriam-webster.com/dictionary/allostasis [accessed 9 June 2019].

causes apart from winter depression (Perreault 2019). Typically, aetiology is outlined, both in childhood and adulthood, practically without any reference to psychosocial conditions.

10.4.2 Is This How the System Works?

I want you to panic.
   Greta Thunberg, environmental activist

What lies under the surface of symptoms? What is under the surface of healthcare systems? Are our healthcare systems built on shallow interpretations of *symptoms,* failing to understand and respect the true *causes* of disease? Can this unpleasant question generally, or only partially, be answered with "yes"? In that case, a fundamental change of concepts would be urgent, a new paradigm needed. I will now present my subjective impression of some parts of healthcare, based on decades of clinical experience in a Norwegian setting.

Experiences that exceed a person's coping capacity result in some kind of damage to health because they provoke harmful stress, emotional pain and pathophysiological consequences. Furthermore, they affect the whole person as well as her or his interactions within ecosystems he or she is involved in, including epigenetic effects.

Every living creature will try to get rid of, regulate or at least soften any form of unpleasant, harmful condition or imbalance. Overeating is only one of many possible ways to achieve relief. We still don't know why certain coping strategies are chosen instead of other alternatives, such as self-harming, drug abuse, and so on. These desperate measures, meant for regulation, can result in a variety of symptoms, some obvious, some hidden and complex.

The obvious symptoms are easily picked up in the healthcare system, interpreted according to rigid professional guidelines and traditions, then categorised in algorithm-based diagnostic systems. One example is the Diagnostic and Statistical Manual of Mental Disorders (DSM), an American concept generally adopted by Western countries. The DSM is overwhelmingly comprehensive, barely leaving any human being without a diagnosis, and it is widely used as a basis for drug manufacturing, financing within healthcare systems, for building up specific departments in healthcare systems and, what is worst: for putting persons into stigmatising categories, often randomly, as their symptoms may fit into many different diagnoses. As a person, you are defined in the light of your diagnosis, and you can be stuck in that narrow prison cell for a long time, deprived of empowerment, deprived of unfolding on your own terms, forced into the world of white coat guidelines.

Generally, it is hard to get rid of a diagnosis once established, regardless of how wrong it might have been. In some aspects, the symptomatic diagnosis is useful, but not as a replacement for the *causal* diagnosis. A patient with a broken leg who has symptoms like *pain* and *not being able to walk* is grateful for symptomatic therapy with analgesic drugs and a wheelchair, but in the long run he will be even more grateful for *causal* therapy: orthopaedic surgery. Persons with obesity are generally denied such *causal* diagnosis, and consequently, deprived of *causal* therapy. The utterly shallow diagnoses keep people chained in a position with impaired quality of life, exploited financially by a number of agencies, including health "care". Causality means not only considering the top of the iceberg, not being stuck in superficial symptoms, but to rather take a look beneath the surface, approaching a deeper understanding.

For the last 80 years or more, the importance of a person's *life story* has been documented. This is where the roots of poor health may often be found, which is a basic premise for cure and enhanced quality of life. Let the patient's life story always be part of the clinical encounter.

We should panic for

- all the people with traumas not receiving adequate help
- the amount of avoidable childhood trauma that our communities fail to prevent
- the healthcare resources being wasted on ineffective or harmful treatment
- the fact that valid knowledge is not implemented in clinical practice
- the fact that our civilisation still stigmatises a large part of the population, meeting them with sarcasm, ruining their self-esteem.

10.5 Outlook

Starting work in the traditional biomedical environment in hospital, I felt the same as expressed by Shorvon and Richardson (1949: 951), "that too little attention had been paid to the emotional factors in the first instance, and the patients had been treated on an organic basis for long periods." This situation may have been caused by the medical funding system that rewards quantity but ignores quality of care. Reasons for this situation may also include the view on causality on which our medical system is based. This view – characterised by lack of attention to the individual and to the interacting factors in the individual's life – also suits business interests too well and contributes to diagnostic and therapeutic restraints. The present volume makes a wonderful contribution to addressing this societal problem.

When, at RSSO, we decided to give much more attention to each patient's life story, setting aside sufficient time for the dialogue, this was a sign of respect, a courtesy to the patients – and I also hoped to create room for new insights and causal diagnosis. The responses from the patients have been positive. They have appreciated being given more time, being listened to, having an opportunity to get rid of shame and to develop deeper insight into how their condition and life story might be interwoven. Many have expressed thankfulness and have been willing to contribute to research. This suits my feeling that we do in fact generate insights in the dialogues.

I have thus decided to try to contribute to research on the severely obese patients from Northern Norway.[3] On 18 October 2018, Martine B Aaseng (patient), professor Linn Getz and the chief executive officer of the Health Administration for Northern Norway (Helse Nord RHF), Lars Vorland, gave an interview on Norwegian television. Getz said that knowledge about the importance of life history is now so well founded that it is time it gets implemented into clinical practice. Vorland responded quite honestly that he had not been aware of this.[4]

As my profession is medical practice, not research, the apparent deficits in diagnosis and treatment of the severely obese were what alerted me to the failure of a system that I had become a part of. But my encounters with researchers Anna Luise Kirkengen, Linn Getz and Rani Lill Anjum helped give me a sound theoretical platform that matches my clinical experience. Four decades after medical school, their work undoubtedly renewed and expanded the basic platform on which my clinical practice is built. This has also had a major impact on my work as GP and at the Norwegian Work and Welfare Authority. It feels right, in every aspect of clinical medicine, to *acknowledge patients' life stories* as essential parts of the *causal* diagnosis that medicine should always strive to achieve. In the clear light of dispositionalism, elaborated in this book, it is now time for changing the game, for a new paradigm in medicine to be accepted and implemented in care, in order to give patients the help they deserve.[5]

References and Further Readings

Bernhardt H (1929) Kapitel I: Zum Problem der Fettleibigkeit. In: Czerny A (ed) Ergebnisse der Inneren Medizin und Kinderheilkunde. Springer, Berlin, pp 1–55

Brewerton TD (2017) Food addiction as a proxy for eating disorder and obesity severity, trauma history, PTSD symptoms, and comorbidity. Eat Weight Disord 22:241–247

Bruch H (1948) Psychological aspects of obesity. J Urban Health 24:73–86

Bruch H, Touraine G (1940) Obesity in childhood; family frame of these children. Psychosom Med 2:141

Danese A, Tan M (2014) Childhood maltreatment and obesity: systematic review and meta-analysis. Mol Psychiat 19:544–554

Felitti VJ (2003) Review of: inscribed bodies: health impact of childhood sexual abuse by Anna Luise Kirkengen. Kluwer Academic, Boston, 2001. Permanente J 7:84

Felitti VJ, Anda RF, Nordenberg D, Williamson DF, Spitz AM, Edwards V et al (1998) Relationship of childhood abuse and household dysfunction to many of the leading causes of death in adults. The Adverse Childhood Experiences (ACE) Study. Am J Prev Med 14:245–258

---

[3] As a board member in the Norwegian Association for the Study of Obesity, I initiated a national conference in Bodø, Norway, in October 2018. Connections between adverse life events and obesity were highlighted in Martine B Aaseng's (patient) testimony and in scientific contributions by researchers Anna Luise Kirkengen, Linn Getz, Rani Lill Anjum, Trine Tetlie Eik-Nes and others.

[4] https://tv.nrk.no/serie/dagsrevyen/201810/NNFA19101818/avspiller; minutes 20.58–23.00 [accessed 9 June 2019].

[5] I would like to thank Kristin Hagen for all her helpful feedback on language, structure and presentation of this chapter.

Garvey WT, Mechanick JI, Brett EM et al (2016) American Association of Clinical Endocrinologists and American College of Endocrinology comprehensive clinical practice guidelines for medical care of patients with obesity. Endocr Pract 22:1–203. https://doi.org/10.4158/EP161365

Hammerlin Y (2014) Når fysisk og psykisk vold blir en reduksjonisme. Tidsskrift for Den norske legeforening 134:1812. https://doi.org/10.4045/tidsskr.14.0939

Hemmingsson E, Johansson K, Reynisdottir S (2014) Effects of childhood abuse on adult obesity: a systematic review and meta-analysis. Obes Rev 15:882–893

Josefsson D, Linge E (2011) Den mörka hemligheten, att lämna det förflutna bakom sig och skapa ett tryggare liv. Natur & Kultur Akademisk, Stockholm

Kierkegaard S, Hong HV, Hong EH (1998) The point of view. Kierkegaard's writings, vol 22. Princeton University Press (first published 1848–49/1859)

Kirkengen AL (2019) En syk kilde [Essay on «The deepest well» by Nadine Bruke Harris]. Tidsskrift for den norske legeforening. https://doi.org/10.4045/tidsskr.18.0622

Lichtwitz L (1923) Über die beziehungen der fettsucht zu psyche und nervensystem. Klin Wochenschr 2:1255–1257. https://doi.org/10.1007/BF01709672

Lynch AI, McGowan E, Zalesin KC (2018) "Take me through the history of your weight": using qualitative interviews to create personalized weight trajectories to understand the development of obesity in patients preparing for bariatric surgery. J Acad Nutr Diet 118:1644–1654

McEwen BS (2012) Brain on stress: how the social environment gets under the skin. Proc Natl Acad Sci U S A 109(Suppl 2):17180–17185

Perreault L (2019) Obesity in adults: etiology and risk factors. UpToDate. www.uptodate.com

Phelan SM, Burgess DJ, Yeazel MW, Hellerstedt WL, Griffin JM, van Ryn M (2015) Impact of weight bias and stigma on quality of care and outcomes for patients with obesity. Obes Rev 16:319–326

Shorvon HJ, Richardson JS (1949) Sudden obesity and psychological trauma. Br Med J 2:951–956

Stensland SO, Thoresen S, Wentzel-Larsen T, Dyb G (2015) Interpersonal violence and overweight in adolescents: the HUNT study. Scand J Public Health 43:18–26

Talbot LS, Maguen S, Epel ES, Metzler TJ, Neylan TC (2013) Posttraumatic stress disorder is associated with emotional eating. J Trauma Stress 26:521–525

Thomas H, Mitchell G, Rich J, Best M (2018) Definition of whole person care in general practice in the English language literature: a systematic review. BMJ Open. https://doi.org/10.1136/bmjopen-2018-023758

Tomasdottir MO, Sigurdsson JA, Petursson H, Kirkengen AL, Krokstad S, McEwen B et al (2015) Self reported childhood difficulties, adult multimorbidity and allostatic load. A cross-sectional analysis of the Norwegian HUNT study. PLoS One 10(6):e0130591

Wang Y, Wu B, Yang H, Song X (2015) The effect of childhood abuse on the risk of adult obesity. Ann Clin Psychiat 27:175–184

Williams O, Annandale E (2018) Obesity, stigma and reflexive embodiment: feeling the 'weight' of expectation. Health, London. https://doi.org/10.1177/1363459318812007

Open Access This chapter is licensed under the terms of the Creative Commons Attribution 4.0 International License (http://creativecommons.org/licenses/by/4.0/), which permits use, sharing, adaptation, distribution and reproduction in any medium or format, as long as you give appropriate credit to the original author(s) and the source, provide a link to the Creative Commons license and indicate if changes were made.

The images or other third party material in this chapter are included in the chapter's Creative Commons license, unless indicated otherwise in a credit line to the material. If material is not included in the chapter's Creative Commons license and your intended use is not permitted by statutory regulation or exceeds the permitted use, you will need to obtain permission directly from the copyright holder.

Chapter 11
Reflections on the Clinician's Role in the Clinical Encounter

Karin Mohn Engebretsen

11.1 Introduction

Several years ago, I decided to examine the philosophical and cultural roots of my therapeutic activities. I was aware of how different ontological perspectives – and in turn methodological choices related to the epistemological question "how do we know" – could affect the therapeutic encounter. There might be some hundred different approaches to psychotherapy but the crucial division between the psychotherapies is not between the "schools" but mainly between what I will refer to as the positivist and the post-positivist or constructivist paradigms. Those years ago, I lacked a clear orientation and became aware that I was vacillating between different methods. I also thought I was able to work without the intention of healing my clients if I just stayed with what was happening in the process. When I realised that, in reality, I did actually have an intention of healing, I decided to explore to see what philosophical theories I might be working from. During this process I wondered about the nature of my underlying motivations for the ontological and epistemological choices I had made in my search for answers to the fundamental questions that are either implicitly or explicitly contained in the way I practice gestalt psychotherapy. Today, psychological theory has become more of a philosophical worldview to me, or a way of thinking and perceiving—more than a taught theory about psychological interventions. In reviewing the path I have followed, I have over time come to know several different traditions in psychotherapeutic practice and, consequently, my opinions about important therapeutic concepts have changed, becoming both extended and refined.

With my desire to heal, came a tendency to see myself as being able to know what was best for my clients. This attitude, I suppose, is still prominent in many clinical encounters. For instance, Cognitive Behavioural Therapy (CBT) is the

K. M. Engebretsen (✉)
Department of Medicine, University of Oslo, Oslo, Norway
e-mail: kme@interdevelop.no

"preferred" methodology seen from a political/governmental point of view. This preference is due to how it fits into the norms, methods and practices of evidence based medicine and the positivist (Humean empiricist) paradigm. Institutionalised norms, methods and practices certainly influence our attitudes when working clinically, and inherent values affect patients in clinical practice and medical care. Although I here use my own experience as a gestalt psychotherapist, I think much of what I say will apply to any encounter where there is a power imbalance as is the case between patients and clinicians. I start from a client-centred approach building on Rogers' (1962) "non-directive" therapy. In this perspective a healthcare system should acknowledge the client as an integrative whole, where the medical issues must be understood not only on the physiological level, but also within a biographical, social and cultural context.

It is equally important to acknowledge the clinician as a person, with everything that he or she brings to the clinical encounter in terms of values, expectations, perspectives and interpretations. How does the clinician influence the encounter with the patient, in positive or negative ways? And how important is it to be aware of one's own role in the clinical encounter?

11.2 Reflections on How Values Affect Clinical Encounters

Historically, gestalt psychotherapy has rejected diagnosis as being depersonalising and anti-therapeutic. This can be seen as a reaction to the dualistic biomedical model, which seems to isolate the issue of psychological suffering as pathological. The DSM (Diagnostic and Statistical Manual of Mental Disorders) is a psychiatric diagnosing tool (American Psychiatric Association 2000) where the diagnostic criteria are, for the most part, based on manifest descriptive psychopathology rather than inferences or criteria from presumed causality or aetiology. The organisational framework by which disorders are grouped into similar clusters are based on shared pathophysiology, genetics, disease risk, and other findings from neuroscience and clinical experience. Being descriptive, it is compatible with how gestalt psychotherapists diagnose clients. In a gestalt perspective, however, psychological suffering is not seen as psychopathology but rather as a creative adjustment to threatening life-experiences and thus, it is based on observations of phenomenology. This includes a focus on body and mind processes, the clients' well-being, character structure and level of emotional development, but also attention to the strength of the therapeutic relationship and stage of treatment. This means that gestalt theory takes into account the total contextual field of the clinical encounter, and thus takes a holistic, non-reductionist and non-dualistic view. As such, the two diagnostic tools differ radically as the DSM does not fully take into consideration the person within his or her context.

When it comes to the diagnostic practices within medicine, I often experience that clients who are referred to me by their general practitioner (GP) are diagnosed with depression. When no physical biomarker is found, the patients' symptoms seem to be attributed to psychopathology and the patients' subjective health complaints are often conveniently reduced to a diagnosis of depression, which may of course be one of their symptoms. This is also often the case for persons who suffer from fatigue and pain related symptoms. The biomedical model and evidence based medicine (EBM) are rooted in the positivist paradigm. Within this paradigm, knowledge is achieved exclusively by what is directly observable and objectively measurable, and little space is left for reflections about subjective factors and underlying mechanisms (Kerry et al. 2012). Being diagnosed as depressed often upsets these clients and they often openly disagree with their GP. The result of such disagreement can result in a lack of trust and worsening of the experienced symptoms.

In clinical interactions, the patient can be addressed as an object, or as a person. Consider the often-used metaphor, that the biomedical model construes the human being as a complex machine. In this machine, dysfunctions might be caused by internal or external harmful factors and the machine is unable to re-establish well-functioning on its own. On this view the person has lost his or her agency and becomes the passive victim of the diseased part subjected to external repair-work.

In contrast, gestalt theory sees the human being as an agent who is in constant interaction with his or her environment, aware of phenomena such as the experience of bodily sensations in response to internal and external interacting factors. Patients are to be recognised as subjects in their own right with their own habitual preferences of behaviour. A clinician cannot truly know what is best for the patient – therefore, it is necessary to give up the desire to be appreciated as some kind of a healer. If not, the patient becomes a means to an end in the clinical process. According to Buber (1965), human interactions can be characterised by a meeting of subjects or a "thingification" of the other. The subject "I", can be seen as part of an I – Thou attitude or the "I" of an I – it attitude. In Buber's terminology "Thou" means "you". Therefore, addressing the client in an I – Thou attitude is a central perspective in relational gestalt psychotherapy and any other person centred practices. Accordingly, I will address two pertinent questions. The first one is: *how important is the clinicians' role within the clinical encounter*?

Gestalt psychotherapy is rooted in an existential-phenomenological world-view. In this world-view, all events are a function of the relationships between multiple interacting forces where no event occurs in isolation (Yontef 1993). If we apply this view to the clinical encounter, any therapeutic process is a function of the relationship between the interacting therapist, client and their common field as a whole. Thus, the field is co-constructed as an integral part of the therapist/client experience, which will have an impact on the possibilities for different outcomes of the process. This means that we can no longer speak of individual growth as "self-development" – in fact it is "self/other development". Additionally, growth of the entire individual/contextual field is only possible if the field has the capacity to adequately support its members. Similarly, in medical practice, any medical treatment might be seen as a function of the relationship between the interacting doctor, patient and the field as a whole, which might be the case in person centred medicine.

In person centred care, the focus is mainly on the patient: how the treatment influences the particular individual, and how the patient responds to the treatment.

First of all, I will highlight the word 'treatment'. How can we understand this term? When you are treating somebody, it is easy to imagine the doctor as providing the patient with something that might be what he or she needs. When the dialogic encounter is understood from a dispositionalist perspective, however, the focus must be on all the participants who are present. As such, the dialogic encounter can be seen as an emergent phenomenon where the client and the therapist are mutual manifestation partners for the outcome of the therapeutic process (see Anjum, Chap. 2, this book). Thus, the dialogic encounter is not simply uncovering the client's experience of her situation, but can be seen as a genuinely interactive process where both the client and the therapist bring themselves in as human beings and thereby influence the encounter reciprocally.

Gestalt psychotherapy is based on the meeting between the therapist and the client as the central healing mode. This means a healing through meeting in reciprocal humanness. In this view it is important to acknowledge the clinician as a person, with everything that he or she brings to the clinical encounter. The development and growth of any healthy self in the field requires a field that includes other healthy selves. We are all inter-dependent and the quality of my life will influence the quality of my environment. Therefore, a relational approach requires careful and consistent observation of all the data in the field including my own processes, values and beliefs as a therapist. This leads us to the second question: *how might the clinician influence the encounter in positive or negative ways?*

In my experience, it is "easy" to handle therapeutic processes individualistically and react as if my existence is separate from my environment – especially when my own self-process is jeopardised. In such cases, how might the encounter be affected?

Being part of the therapeutic process, I am not only engaged as a supportive ground, but as a co-participant as well. When "acting" as a tool in the therapy process, the therapist might be drawn into phenomena such as transference and countertransference (Rycroft 1979). These two notions can be understood as the process by which feelings, behaviours or attitudes of clients and therapists that belong in the past, are transferred to the therapy participants in real time. Transference is our unconscious activity that is shaped by normal preverbal perceptions of self and other, which organise our subjective universe (London 1985). Client and therapist actively co-create the shared perceptual field of the therapeutic relationship. Therefore, from a dialogic perspective, the client's processes of transferring cannot be interpreted as emanating from the client in isolation, but must be seen as emerging as part of an inter-subjective relational system (Hycner 1991). Countertransference refers to the therapist's feeling towards the client, in response to the projected transference.

When the therapist is unconsciously drawn into transference processes, the outcome of the therapeutic relationship might in best case be ruptured and in worst case be quite the opposite of therapeutic. In my experience, relating objectively to phenomenological data constitutes a major challenge, even for well-trained psychotherapists. Therefore, I see supervision as crucial when working dialogically. I have experienced meeting clients who are sensitive to what they experience as personal critique. Their emotional reaction might be due to shame proneness. Some of these clients seem to have been severely shamed by previous therapists, before seeing me. The presentation of the work I did with one of my clients later in this text shows how easily this can happen, and also reminds us of the importance of being consciously aware of the therapeutic process. The attentive attitude requires humility and explicitly promotes respect and appreciation of differences.

I am aware of how contextual conditions often change the way I work. With clients who are more psychotic, I work more analytically – just being there, holding the boundaries, not intervening, challenging or contacting. With healthier clients, I work more dialogically. These choices mirror the clients' level of emotional development and character structure as well as the strength of the therapeutic alliance. I am also aware that there might be a difference between theory and practice – how my values might change in practice. In one-to-one settings I am gentle and soft; in groups I am often more robust and challenging. The danger of not taking differences between people seriously enough is constant. In my experience, it is exactly the art of relational psychotherapy to bring the differences into awareness. And in the dialogical encounter differences related to values or attitudes are to be appreciated – not diluted or combatted.

In what follows I present a snapshot of the psychotherapy I enact. By presenting the work I did with Marie over several months, I will illustrate how my work embodies the theory. This illustration shows how the phenomenology I enact fits into the dispositionalist paradigm. I start with a presentation of how I experienced the initial meetings with Marie and my reflections on what she might need. Then I describe parts of the work we did together.

11.3 The Work I Did with Marie

11.3.1 Presentation of the Client

Marie[1] was a 45 year old married woman with two grown up children. Her husband was a chief executive officer in a large multinational company and travelled a lot. Due to this she spent most of her time on her own. Until recently Marie was a manager in a small company, but lost her job last winter. This made her feel lonely and lost. She told me about her happy childhood; her mum and dad and her sister who was 3 years younger, whom she adored. The family spent a lot of time together, either alone or with friends. Both her parents were dead, her father died when she was 20, her mother 2 years ago. After the death of her mother, she did not have much contact with her sister, although they lived in the same part of the town. Before she began seeing me, she had been ill for a month. Her referral to me was via her GP, who had diagnosed her as depressed. Her reason for wanting therapy

---

[1] 'Marie' is a fictive name. Some details have been altered for the purpose of de-identification according to ethical and legal standards and written consent to use her story has been obtained.

was that she needed some help to fix her life. She presented with issues of anxiety and panic attacks, feeling isolated from others, especially her husband. Her GP wanted to give her antidepressants, but she refused because she did not see herself as depressed. We agreed to work for a 6 week assessment period to determine whether we were able to work together or not and to review our work after that on an open ended contract.

The first time I saw Marie, I was struck by her attractiveness; she was tall, slender, and well dressed, with a determined stride. Her long dark hair framed her deep blue eyes. Her voice was rich, and sensual. With the pitch of her head forward, her eyes were often cast downward and seldom met mine. Sometimes, when she did look in my direction, she glared. This look of camouflaged contempt made me feel tense. Marie's facial expressions were endless. She commented on everything with a wince, as if every feeling that passed through her body was expressed only by the muscles of her face. When she became anxious in the session, this tendency was especially evident.

My initial reaction to Marie was curiosity and I felt warm and concerned towards her. I experienced her as being extremely bright, demonstrated by the way she presented herself and her use of vocabulary. Marie's connection to her mother changed at the age of three, when her sister was born, and she was sent to her mother's sister for some weeks. Afterwards she became dad's girl. When telling me about her father's death, she teared up, but seemed to be unable to relate to the emotional situation that she obviously experienced. She turned away from me, silently sitting there for a while – then she laughed. I often experience such reactions, which I see as a normal human ability that allows us to put off dealing with emotions until we feel able to address it. When she laughed, I was aware of feeling irritated and when reflecting on why, I became aware of one of my personal assumptions that I might have acquired without full awareness of its purpose: "It's silly to laugh". She described her relationship with her mother as ambivalent, never knowing where she was in relation to her. Marie told me that her mother must have been depressed – she could be silent and withdrawn for days, not addressing anybody. I was struck by how some aspects of her history paralleled my own. I also felt adored by my father and abandoned by my mother, which alerted me to the possibility of transference/countertransference processes. And, I was aware that I could easily be drawn into over-identifying with her. I knew at this early stage that I would need to discuss our relationship with my supervisor to be able to bracket off my own emotional baggage.

11.3.2 Presenting Problems

Marie came to therapy with difficulties especially in her relationship with her husband. The slightest misunderstanding between herself and others left her with feelings of abandonment and deep loss. Whenever she spoke about an emotion such as her fear of being abandoned, she would immediately discount it with statements such as, "But I know that's crazy, because I should not feel that way!" In this regard she was extremely critical of the vulnerable aspects of herself. She was also highly critical of others, an aspect of herself which she joked about by stating that it was due to her superiority complex.

11.3.3 Diagnosis

From a gestalt perspective I think it is fruitful to understand Marie's process historically as a "creative" adaptation to her life situation. This adaptation can be seen as the relationship between Marie and her environment, in which she takes responsibility for creating conditions to take care of her own well-being. Thus, I see diagnosis as a descriptive statement that articulates what I notice in the present, which informs me of how I might be able to help my client.

Gestalt psychotherapy embraces Merleau-Ponty's holistic view of the human being, conceptualised as existing in continuing interplay in the "organism-environment field" (1945/1962). The organism-environment field can be understood as a systematic web of relationships, which consists of a totality of mutually influencing forces that together form a unified interactive whole. Out of this intersubjective field, "figures" emerge. The configuration of a figure against a ground displays the meaning, and meaning is achieved only through relations in the field. Thus, the relationship between the ground of the field and the figures that emerge is what gives meaning to the whole. To perceive and be aware of an emerging figure is the act of contact.

The idea of "unfinished business" is a core notion in the gestalt approach to explain how the act of contact might be interrupted. This notion refers to a tendency of the organism to complete any situation that is experienced as unfinished (Perls et al. 1951). For example, when Marie was not able to get her needs met, some specific contact episodes emerged between us at the contact boundary (Clarkson 1989). I became aware of some aspects of her behaviour, which stood out from the context like figures against a ground. These figures became interesting as a source of further exploration when I observed them as a pattern over time and across situations. Contact boundary disturbances do not refer to psychopathology, but to a disruption in the clear awareness and organismic flow between self and other, which can be either healthy or pathological. Very early on in the encounter with Marie, I became aware of the transference that was taking place between us. I reflected on whom I might represent to Marie – when I felt warm towards her, I would be her father, and when I felt irritated and critical, I would be her mother. So, when I was feeling irritated and critical towards her, she might have been unaware of conveying her feelings by giving me as the therapist an experience of how she feels, rather than by articulating. Thus, there was a possibility that I could end up behaving towards Marie like her mother did.

After this brief presentation of my initial contact with Marie, I now turn to the therapeutic process and present some of these interruptions to contact, and how I work dialogically.

11.3.4 The I-Thou Process

To illustrate how I worked with my client, I will list four discernible phases in the "I-Thou" perspective (Buber 1965). These phases are (1) "Exploring self; an it – it attitude", (2) "Becoming aware of the therapist's presence; an I – it attitude", (3) "Struggling with abandonment depression; an I – Thou attitude", and (4) "Moments of mutual satisfaction; a Thou – Thou attitude". These four phases show how Marie's and also my own ability to integrate personality aspects that previously were not fully "owned" resulted in an emergent phenomenon, that is, the therapeutic outcome.

Phase 1: Exploring Self; an It – It Attitude. When Marie entered therapy she talked about herself and objectified both herself and me as a therapist and asked, "how can you fix me?". Marie rapidly established an "idealising" transference towards me, which can be understood as Marie's unconscious recognition of some of her mother's traits in me, and then started acting out how she previously idealised her mother. When I became aware of the idealising transference she projected onto me, and the immediate impact it had, I realised how flattered I felt. I was able to see that I was not fully present to her as another person. However, the loving attitude I felt towards her, and the mirroring I did during this phase, was an authentic desire to nurture and "mother" her.

My main goal for this phase (6 weeks of therapy) was to build a therapeutic alliance. I focused primarily on building a trusting relationship, and therefore I was initially and primarily concerned with "confirming" her (Buber 1965). By confirming in this context, I mean accepting not only what Marie is aware of, but also aspects of her existence that are denied, e.g. confirmation of the person in her fullest potential. I started to practise "inclusion" (Buber 1965) with her. By this I mean closely listening to both verbal and nonverbal communication, carefully giving her phenomenological feedback to raise her awareness of herself. In this situation I address her as a "person" – not as an "object". Intuitively I felt she was very sensitive to anything I did that she could interpret as being a rejection. However, instead of telling me directly when she felt ignored or insulted by me, she would get a certain withdrawn and contemptuous look on her face that I came to recognise. When I addressed this phenomenologically, she would comment back to me with a wince, obviously feeling misunderstood and attacked. She was not interested at this point in insight about herself, because she was convinced that all insight would simply lead to criticism. I was imagining that her self-esteem was very fragile and instead of exposing her insecure self, she presented a "false", defensive self to me. This imagining must be distinguished from empathy, which leaves out one's own side as a therapist. To be able to practice inclusion the therapist needs to be able, as much as is humanly possible, to attempt to experience what the client is experiencing, feeling, thinking or knowing from her side of the dialogue, as well as meeting her authentically and honestly as part of practicing inclusion with her.

Phase 2: Becoming Aware of Therapist's Presence; an I – It Attitude. Previous sessions had taught me that experiments had triggered resistance and would be seen as criticism of Marie's behaviour. For example, Marie suddenly stopped talking in terms of herself and switched from saying "I" to saying "we" without any apparent awareness. Instead of saying: "why don't you try an experiment and say 'I' instead?" I would rather say: "I was wondering if what you were talking about suddenly felt too painful to continue talking of in terms of yourself?" This response made it easier for Marie to explore her painful feelings and helped her to stay in contact with me, and increased her self-awareness without triggering unbearable anxiety.

Marie had introjected her mother's self-image and idealised her mother in order to maintain any sense of having an ordered, loving family. An "introject" may be seen as accepted personal habits acquired without full awareness of their meaning and purpose (Perls et al. 1951/1998). When she described what she was aware of when she attempted to make contact with her husband, she became more aware of her "impasse". Here, the impasse can be understood as how Marie acted out the experience of seeing herself as a dis-empowered object (Newirth 1995). She imagined that her husband was much too busy to want contact with her, and the conversation just stopped. This experience left her frustrated, lonely and longing for connection. I asked her to describe her experience of longing, which she experienced as a vulnerable and lonely feeling in her stomach. She added that she feared rejection and quickly stated that her mother had never accepted her husband – and he was not worth connecting to anyway. I encouraged her to stay with this feeling of criticism, and we explored further the frustration that emerged in her. When she was able to disclose more of her feelings of insecurity and low self-esteem, she was able to ask for more support. At this point, she was able to take in the support I offered her when raising her awareness of how her attempts to deal with her vulnerability by being critical towards herself only left her feeling more frustrated and distressed. Gradually she became able to honour herself.

During this phase I experienced that she started to sense her feelings of anger, being aware of her need to express herself, to mobilise her energy, and finally to vent her feelings towards her mum.

Phase 3: Struggling with Abandonment Depression; an I – Thou Attitude. Once this trusting and accepting relationship was established, the next phase began, where some of Marie's problems related to interactions with others were further explored. There were more moments in which I – Thou encounters occurred, than there were moments in which I – it encounters occurred. As Marie started to see me as a person I became more important to her. This was when the therapeutic relationship was challenged.

11.3.5 Key Episode 1

Marie started to project the anger outwards, which she had previously controlled, and became very critical towards me. When I heard her stating: "no one can be relied upon", I was aware that this was similar to one of my own introjects and I recognised the wounds from feeling rejected by my mother. It was therefore important for me to "bracket off" my own emotional reactions in order to be available to explore and to fully understand how Marie made meaning out of this statement. In this context "bracket off" means that I held some of my own concerns in abeyance in favour of attending to what was going on when interacting with Marie. I tried to meet her honestly and authentically. I did, however, feel wiped out by her and knew that this was something I would have to explore in supervision. Initially, I was able to stay in contact with her, but I was aware of feeling induced to behaving towards her like my mother did towards me. And because of the sensed similarity between Marie's mother and my own, I was particularly vulnerable to this induction.

When she started to reveal her dependent, needy side I was aware of feeling irritated and angry with her. I had, in a sense, fulfilled what she expected – being rejected was what she really feared. She exhibited the anger she previously had controlled and talked to me with sarcasm and contempt, when the narcissistic and maternal "supplies" she was seeking from me were withheld. I thought in that moment and subsequently that Marie had benefited from my firm, withholding posture and that I had managed to resist her seductiveness. I was however left feeling uncomfortable about how harsh I had been and continued to be over the next two sessions. I sensed that Marie was withdrawing. We both had reached an impasse. I struggled a lot and felt dreadful until I had discussed what had happened and worked through the process with my supervisor.

Phase 4: Moments of Mutual Satisfaction; a Thou – Thou Attitude. In this last phase Marie was able to practice inclusion with me, which means that she was able to stay in the present moment, meeting me honestly and authentically. This happened after I decided to disclose the pain I had felt and explained to Marie what I thought had happened in the process.

11.3.6 Key Episode 2

I was able to disclose my humanness to her and say I was sorry for the mistake I had made after being able to non-defensively own parts of my own history and been able to heal old wounds. When doing so, the contact between us was paradoxically re-established. I felt I was risking a lot, but in this moment of Thou – Thou mutuality, being authentically present, I felt grace and deep satisfaction. By authentically disclosing myself as who I am, something changed.

Over the course of this work, Marie started to see herself differently. There has been a shift in her attitude towards herself. Instead of seeing herself as the bad part of the mother-daughter relationship, she has started to see that it was not all her fault. In this process, she has started to grieve for the mother who was not there for her. Marie reported that the panic attacks had not occurred since Key Episode 2. She is now able to feel more satisfaction in her life, and her relationship with her husband has deepened.

11.4 Reflections

I see philosophy as the way I hypothesise about everything in my life and the essence of how I work as a psychotherapist and researcher. Philosophy also helps deconstruct being social; "how do I connect with what is?". This fundamental question is an inquiry into what collective experience is possible. On the contrary, to treat knowledge as intellectual, prescribed and something we are taught, is to forget its social and interpretative nature. Thus, this question is of concern to all of us, and not so much in terms of "I", more so in relation to "we".

In this short text I have reflected on the clinicians' role and how we as clinicians might influence the outcome of the encounter. I am intrigued by the relationship between the therapist and the client, as well as the relationship between the client/therapist and "significant others" in the clients' lives. This relational aspect is the concrete basis for how I work clinically. The ontological stance we take, either consciously or unconsciously, will influence our norms and methods and, in turn, the way we practice. I hope that this text will stimulate further reflection and discussion on the clinicians' role in the clinical encounter.

References and Further Readings

American Psychiatric Association (2000) Diagnostic and statistical manual of mental disorders, 4th edn. APA, Washington, DC
Buber M (1965) The knowledge of man. Harper & Row, New York
Clarkson P (1989) Gestalt counselling in action. SAGE Publications, London
Hycner R (1991) Between person and person: toward a dialogical psychotherapy. The Gestalt Journal Press, Gouldsboro
Kerry R, Eriksen TE, Li SAN, Mumford S, Anjum RL (2012) Causation and evidence-based practice: an ontological view. J Eval Clin Pract 18:1006–1012
London N (1985) An appraisal of self psychology. Int J Psychoanal 61:237–248
Merleau-Ponty M (1945/1962) Phenomenology of perception. Routledge, London/New York
Newirth J (1995) Impasses in the psychoanalytic relationship. Sess: Psychother Pract 1(1):73–80
Perls F, Hefferline RF, Goodman P (1951/1998) Gestalt therapy, excitement and growth in the human personality. The Gestalt Journal Press, Gouldsboro
Rogers CR (1962) On becoming a person. A therapist's view of psychotherapy. Houghton Mifflin, New York
Rycroft C (1972/1979) A critical dictionary of psychoanalysis. Penguin Books, Hammersworth
Yontef G (1993) Awareness, dialogue and process: essays on gestalt therapy. The Gestalt Journal Press, Highland

Open Access This chapter is licensed under the terms of the Creative Commons Attribution 4.0 International License (http://creativecommons.org/licenses/by/4.0/), which permits use, sharing, adaptation, distribution and reproduction in any medium or format, as long as you give appropriate credit to the original author(s) and the source, provide a link to the Creative Commons license and indicate if changes were made.

The images or other third party material in this chapter are included in the chapter's Creative Commons license, unless indicated otherwise in a credit line to the material. If material is not included in the chapter's Creative Commons license and your intended use is not permitted by statutory regulation or exceeds the permitted use, you will need to obtain permission directly from the copyright holder.

Chapter 12
The Relevance of Dispositionalism for Psychotherapy and Psychotherapy Research

Tobias Gustum Lindstad

12.1 Introductory Preface

Being educated as a clinical psychologist, I am grateful for having had the opportunity to work with unforgettable patients and colleagues. Yet, having worked within both secondary and primary mental health care services for the last 16 years, something has persistently felt wrong with my working conditions. Having shaken off my worry that this feeling is a symptom of something being wrong with me, I have become convinced that a significant part of the problem is the scientific paradigm of psychology, which does not only set the conditions for psychotherapy research, but also for clinical perspectives of relevance for any health professional working in mental health care.

Through the history of psychology, psychologists have sought to legitimize their discipline as a science by differentiating it from the humanities. However, as such, they have unfortunately put more faith in the processing of computers working inductively on accumulations of empirical data, than to their natural abilities to scrutinize their assertions via thorough reflection and in critical dialogue. One particularly detrimental aspect of this paradigm has been the predominant presumption that statistically supported empirical experiments in the form of randomized controlled trials (RCTs) are needed for clarifying the causal effects of psychotherapy. However, this idea is not so scientific as it is bad philosophy. Not only does it imply questionable conceptions of causality, but it also neglects many natural characteristics of being a person.

Moreover, it does not help much that results of RCTs have been used as a bureaucratic remote control by governmental health authorities wanting to assure the quality of health services from the outside (that is, without taking part in the process of providing the service). On the contrary, this has made clinicians walk around on

T. G. Lindstad (✉)
Independent researcher and Clinical Psychologist in Private Practice, Åros, Asker, Norway

© The Author(s) 2020
R. L. Anjum et al. (eds.), *Rethinking Causality, Complexity and Evidence for the Unique Patient*, https://doi.org/10.1007/978-3-030-41239-5_12

their tiptoes needlessly worrying about whether they conform sufficiently to standardized procedures thought to have had results on an average level. This runs the risk of dehumanizing mental health care services by not taking sufficiently into account the context-bound complexities of clinical encounters and by being an obstacle for a sufficient focus on the unique needs of individual patients.

Accordingly, though this book promotes an account of causality relevant for all health sciences and professions, the focus of this chapter is on how dispositionalism may improve upon the foundations of clinical psychology, psychotherapy research and mental health care services. This is important not only for psychologists, but for all clinicians providing psychotherapy or related services (nurses, physiotherapists, psychiatrists, physicians, social workers etc). Not only will the recent advancements of dispositionalism (see Anjum, Chap. 2, this book) provide resources for a refreshingly new foundation for psychological science and psychotherapy research, but also for more humane mental health care services.

12.2 Misleading Statement on Evidence Based Psychological Practice

Philosophers have often considered it a virtue to be informed by psychology. Psychologists, however, though they may admit philosophical inspiration, have only rarely declared philosophy as relevant for improving psychological science and practice. Accordingly, the widely acknowledged statement on evidence based psychological practice provided by the American Psychological Association (APA 2006: 273–4) puts a one-sided emphasis on empirical research and neglects the relevance of philosophical reflection. As such it simply upholds that statistically supported empirical experiments in form of RCTs are the standard for drawing causal inferences about the effects of psychotherapy. However, as the APA-statement is a significant supplier of terms for evidence based mental healthcare services in general, this emphasis of the statement is beset with difficulties. Not only is the prevailing presumption that RCTs is the standard for clarifying causal relations part of a questionable Medical Model of psychotherapy (not to be confused with the biomedical model discussed in Chap. 5 of this book), but it also hinges on dubious conceptions of causality inherited from Hume and is therefore not something one should take uncritically for granted (see Anjum, Chap. 2, this book).

To be fair, to some extent the APA-statement also seems to have relevantly rectified the medical model. Not only does it provide a broader definition of evidence based psychological practice as the integration of the best available research evidence with clinical expertise in the context of patient characteristics, but it also approves of multiple types of empirical research evidence, not only RCTs. However, though these are important steps in the right direction, the understanding of evidence based mental health care must be broadened so as to include philosophical and theoretical reflection, and moreover, and accordingly, the predominant idea that

RCTs are the best way of drawing causal inferences for the single patient must be abandoned.

Notice, that though my call for change is critical, it does not have destructive aims. Rather, it is part of a constructive counter-reaction that seeks to liberate the health sciences from the detrimental impacts of Humean conceptions of causality and to pave the way for more apt alternatives. Fortunately, the Humean regularity view, along with its descendent counterfactual and difference-making accounts, are not the only accounts of causality available. As such, the recent philosophical advancements of dispositionalism (e.g. Mumford and Anjum 2011; Anjum and Mumford 2018a) do not only provide relevant resources for bringing psychology out of its dead ends, but they may also breathe new life into pertinent alternatives that have been unheeded. What is at stake is nothing less than the understanding of what relevant psychotherapeutic competency is, how it may develop, and how the quality of mental health care services may be assured.

12.3 Questioning the Medical Model

Despite attempts to overthrow its predominance (Wampold and Imel 2015; Duncan et al. 2010) the Medical model still thrives as the following tripartite set of presumptions:

(i) RCTs are the best way to clarify causal effects.
(ii) Evidence based psychotherapy depends upon the clarification of causal effects of *specific* treatment interventions and methods (often derived from *specific* psychotherapy-models) on *specific* disorders categorized according to *specific* symptoms frequently observed together.
(iii) The implementation of such empirically supported treatment methods (so-called ESTs) is what evidence based psychotherapy should amount to (cf. Chambless and Hollon 1998).

As medical research and practice are more varied than what these claims amount to, the Medical model could perhaps be better called "the Pill Model". This fits the idea in question that psychotherapy should be studied and understood in the same way as drugs. Though this model has never been generally accepted, the idea it represents has been very much alive among influential scholars. E.g. Kennair et al. (2002: 9) have claimed that "the major conclusion after years of research into the effects of psychotherapy is that certain interventions work for specific disorders", and that though "there are variations between humans, … there also is a relatively uniform human nature [which] means that interventions that work on large groups of humans will probably work for random individuals". Accordingly, proponents of the Medical model have not only argued that psychological treatments require theories of causal relations and mechanisms of change, but they have also misconstrued psychotherapy as the systematic use of psychological knowledge in such a way that it leads to expected change with statistical probability. Accordingly,

```
pyschotherapist₁ ⟶  Statistically probable           Specific symptoms    ⟵ person₁
pyschotherapist₂ ⟶  difference maker:                of specific disorder ⟵ person₂
                    Specific interventions of
pyschotherapist₃ ⟶  specific treatments derived                           ⟵ person₃
...                 from specific treatment                               ...
pyschotherapistₙ ⟶  models and perspectives                               ⟵ personₙ
```

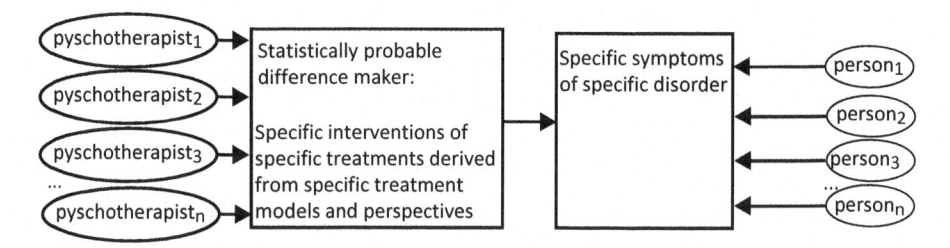

Fig. 12.1 The Medical model

Fig. 12.1 illustrates how the predominant Medical model portrays any psychotherapist as complying with one or more empirically supported treatments which purportedly makes a statistically probable difference on specific symptoms of a specific disorder that are allegedly shared by various patients.

12.4 The Challenge from Dodo-Birds and Meaning-Makers

As mentioned, however, the Medical model has not remained unchallenged, and roughly put, seminal critiques have come from two partly overlapping groups of scholars; to keep them apart I will henceforth call them the *Dodo-birds* and the *Meanings-makers*. The Dodo-birds have long argued that "psychotherapy does not work in the same way as medicine" (Duncan et al. 2010: 28), and that RCT-research has "failed to find a scintilla of evidence that any specific ingredient is necessary for therapeutic change" (ibid. 33). Moreover, they have claimed that the empirical research originally conducted in accordance with the Medical model actually produces evidence that supports an alternative contextual model on which the methodological and technical aspects of therapy processes cannot be studied as isolated from the relational context in which they are part (Wampold and Imel 2015). Thus, rather than continuing the search for the one and only miracle cure, the Dodo-birds have emphasized the so-called Dodo-bird-verdict first uttered by the fabulous dodo-bird in Lewis Carroll's tale of Alice in Wonderland: "Everybody has won, and everyone must have prizes". Accordingly, the Dodo-birds have claimed that there are no evidential statistical differences between treatment models and have stressed the importance of common factors purportedly transcending any specific treatment method, such as client-, therapist- and alliance factors, as well as factors external to the therapy, and that psychotherapy processes must be "evidence based one client at a time" (Duncan et al. 2010: 39–40).

Several Meaning-makers have agreed with these conclusions. However, their arguments have also cut more deeply by pointing out that the research-design of RCTs ignores fundamental aspects of human beings, such as responsiveness and irreversible uniqueness (e.g. Stiles 2009; Smedslund 2009). So far, so good. However, often inspired by phenomenology or Wittgenstein, the Meaning-makers have taken the even more radical step of rejecting any psychological relevance of causal explanation as they think the notions of causality and human meaning-making belong to different explanatory domains (Brinkmann 2011; Harré 2002; Smedslund 2012a). I principally agree with both groups of critics. However, I also argue that they have not taken the bull by its horns, and that acknowledging the relevance of dispositionalism will provide the relevant improvement on their arguments. What is wrong is not to emphasize causation, but rather the Humean ideas about causality that nourish the Medical model.

12.5 The Philosophical Bias of the Medical Model

Whenever one emphasizes some research methods rather than others, one takes for granted that the phenomena one wants to study may have properties that make one's choice appropriate. As such presumptions might be wrong, it ought to be considered a scientific virtue to make one's presumptions about what one studies available for scrutiny. As such, the abovementioned claim of Kennair et al. (2002: 9) admirably illustrates that the critics of the Medical model have not been attacking a strawman. The claim of Kennair et al. that "there … is a relatively uniform human nature [which] means that interventions that work on large groups of humans will probably work for random individuals" does not only rest on ill-founded Humean presumptions about causality, but it also errs seriously on behalf of human nature.

However, as argued by several Meaning-makers (e.g. Smedslund 2009; Stiles, 2009), physiological research is not needed to demonstrate the relevant variability. There is more to human beings than mere measurable existence: Unless we are in a coma, we are compulsively meaning-making persons for whom something exists, and once something has been experienced, this is irreversibly so. As we also unavoidably attach meanings to the world from never identical contexts, the complete sets of our experiences become inevitably unique. And moreover, we are continuously susceptible to change by attaching new meanings to our experiences. Thus, the ways we make sense of things are bound to evolve in unique ways within unique contexts. Hence, though this does not prevent similar experiences, we cannot take for granted that persons will react in the same or similar ways to the same event or similar events.

Most textbooks on psychological research methods have acknowledged that persons have characteristics that make them difficult to experiment with. Nevertheless, apparently because it is held that these difficulties can be circumvented via statistics, the view still prevails that RCTs are the best way to uncover causal relationships (e.g. Hollon 2006). Thus, regular causal effects are standardly not sought for on an individual level, but rather on an average level (APA 2006). By randomly assigning a high number of persons to groups subjected to different conditions – for instance, offering some persons psychotherapy while others not – and estimating subsequent statistically significant differences between the average scores of the groups, one may conclude that the differences have been caused by the psychotherapy. The viability of this conclusion depends on the two groups being similar in all other relevant respects, and it is held that this is taken care of by the randomization procedure. However, though characteristics that are possible for people to share (e.g. height) may spread evenly in large-sized random groups, unique characteristics (e.g. memories) cannot. Thus, as no randomization procedure can prevent unique experiences (e.g. memories) from being influential, we cannot take for granted that groups are relevantly similar without having thoroughly considered the unique experiences of the individuals involved. Also, simply increasing the number of persons in the hope that the statistical law of large numbers will apply will not help, as increasing the number of unique aspects does not necessarily make the groups any more similar, but would rather increase the numbers of influential factors that ought to be taken into account for understanding why the results occur. In other words, if we do not get a sufficiently thick understanding of the unique experiences of the individuals involved, information about aspects that inevitably influence the results are missed (Smedslund 2009; Stiles 2009). Hence, RCTs risk throwing the baby out with the bathwater. RCTs can only tell us *that* psychotherapy has had such and such results on a mean level, but neither *why* these results occurred, nor *how* and *whether* similar results are attainable now or in the future.

12.6 Dodo-Birds Must Take the Bull by Its Horns

Unfortunately, the above critique of the Medical model has been largely unheeded. Admittedly, the APA-statement does offer related rectifications by providing the somewhat broader definition of evidence based psychological practice mentioned above, as well as by approving of other empirical research designs than RCTs, such as the process-outcome studies often emphasized by the Dodo-birds (Orlinsky et al. 2004). However, by upholding RCT's as the standard for drawing causal inferences about the effects of psychotherapy and even declaring that barriers to this kind of research should be identified and addressed (APA 2006: 274), the statement uncritically makes the same ontological commitment to a regularity view of causality as do proponents of the Medical model. As such, rather than explicitly acknowledging the contextual model as an apt meta-understanding to replace the Medical model, the APA-statement does not only contribute to maintain the flawed idea that replicated RCTs are the best research design that we have, but at worst it inspires proponents of the Medical model to uphold the contextual model as just another specific form of treatment that must be evaluated by RCTs (e.g. Crits-Christoph et al. 2014).

However, if the prevailing emphasis on RCTs relies upon a faulty understanding of causality, the argument of the Dodo-birds that evidence from RCTs does not support the Medical model does not cut deep enough, as it has not explained why this is so. That is, pointing out that the statistical evidence indicates that the presumption of the Medical model that specific treatments work for specific disorders is wrong, does not explain why it is wrong. However, as the arguments of the

Dodo-birds wave in the relevant direction, it is to be hoped that they are also prone to open their eyes and take the bull by its horns. Indeed, the reason Duncan et al. (2010: 36) are right that "clients are not dependent variables on which independent variables operate [but] agentive beings who are effective forces in the complex of causal events", is that the Humean conception of causality is wrong and dispositionalism right.

12.7 Meaning-Makers Must Target the Right Enemy

Where the Dodo-birds have not yet taken the bull by its horns, the Meaning makers have overshot their target. If there was no alternative to the prevailing Humean conceptions of causality, the Meaning-makers would have been right to deem causal explanation as relevant only for the natural sciences while inapt for psychology (Smedslund 2012a; Harré 2002; Brinkmann 2011). However, as alternative accounts of causality do exist, their rejection of the relevance of causal explanation amounts to no less than overkill.

It should be noticed, however, that this rejection stands in a long tradition of scholars inspired by Dilthey's distinction between explaining via causal covering laws (*Erklärung*) and understanding agents' points of view (*Verstehen*) (Harré 1999; Smedslund 2009). This distinction is indeed relevant, as the very point of RCTs – to compare the outcome of therapy with no therapy – is clearly related to the so-called covering law model, at least in Hempel's version inspired by Hume. According to this model, scientific explanations should reveal the regular antecedent conditions without which something would not happen by appealing to empirical correlations of one type of event (outcome) with another type of event cited as its cause (intervention). However, that this kind of explanation by referring to empirical correlations (covering laws) is inadequate for understanding human meaning-making, entails neither that meaning-making and causality belong to mutually exclusive explanatory domains (cf. Smedslund 2012a; Harré 2002; Brinkmann 2011) nor that it is required to think about human action in ways that go beyond models of causality (Valsiner 2014). Thus, though Valsiner is right that sticking to search for linear causality has led psychology to ignore the possibility of alternative accounts of causality, Valsiner and Brinkmann (2016) are wrong that any human sign-regulatory system is a catalysed, *rather than* a causal system (my italics). Rather, in line with Valsiner's (2017) more recent statement, talk about causality must take *a new form*, and as such, the recent advancements of dispositionalism (Mumford and Anjum 2011; Anjum and Mumford 2018a) seem related to Valsiner's suggestions to emphasize catalytic conditions and mutually beneficial relations rather than relations between independent and dependent variables.

To sum up: both Dodo-birds and Meaning-makers have compellingly argued against the second and the third presumption of the Medical model described in Sect. 12.2, but to no avail. However, as an apt dispositionalist alternative to Humeanism exists, also the first presumption of the Medical model must be abandoned: Not only is it wrong that evidence based psychotherapy is equal to empirically supported treatments, and senseless that specific kinds of outcomes must be repeatedly found to follow from specific kinds of interventions, but more fundamentally, it is also wrong that RCT's are always needed for drawing causal conclusions.

12.8 Humeanism Must Be Replaced by Dispositionalism

The recent advancements of dispositionalism relate to the resurgent philosophical interest in understanding the relevance of dispositional properties for causality (e.g. Groff and Greco 2013). As such properties are also often called causal powers, dispositionalism has often been presented as influenced by Rom Harré's seminal efforts (with Madden 1975) to replace the Humean regularity theory of causality with an account that revivifies the notion of causal powers. However, though Harré's contributions to discursive psychology (e.g. with Gillett 1994) are relatively well known to psychologists, his contributions to the philosophy of causality have been largely unheeded. Probably, this is not so much because Harré (2002) was reluctant about the relevance of causal explanation for psychology, but because psychologists are rarely encouraged to be philosophically informed. Yet, the relevance of causal powers has long been discussed in relation to the social sciences (Groff 2008), and with the recent advancements of dispositionalism their pertinence for medicine (Anjum 2016) have been demonstrated. However, pace Harré, these advancements are no less relevant for psychology. The problem is not that psychologists have wanted to discover causal relations, but that their Humean conceptions have been misleading.

A growing number of philosophers now regard the long prevailing Humean regularity theory as no more than a standard against which to contrast and develop more refined accounts. Though the extent to which Hume himself really held the view is controversial, there is no doubt that he put both it and the related counterfactual difference-making account on the table. On the first view, causal relations (infamously exemplified by colliding billiard balls) consist in no more than that events of one kind can be observed as regularly conjoining or following events of another kind. On the latter view, causes are events without which their effects would not happen. However, on both views, causal relations are neither governed by any necessities nor by any dispositional properties, or if they were, we could simply not know. Notice, that as these Humean conceptions imply that causal links must either be demonstrated by statistical evidence of correlation and/or by comparing the average outcome of exposure by stimuli with the average outcome of no exposure, they fit with the Medical model's emphasis on RCTs like a glove. Probably, the Medical model's predominance is also due to the continued influence of Hempel's covering law model, in which scientific explanations should reveal the regular antecedent conditions, as well as the empirically observed general laws, without which something would not happen (Groff 2011).

However, as pointed out by prominent advocates of dispositionalism, scientific observations of regularities that cannot be prevented by any means hardly exist (Mumford and Anjum 2011). Probably, this is why the number of citations of general laws in the scientific psychological literature have decreased (Teigen 2002). Yet, the predominant scientific paradigm of psychology still emphasizes research methods developed for the empirical discovery of regular causal relations and statistical differences. However, in line with dispositionalism it should be noted that methodological questions of how to discover causal links should not be confused with ontological questions of what causality is (Anjum and Mumford 2018b). Thus, though differences between group averages may indicate that relevant causal relations exist, causality is not in itself a statistical phenomenon. Without further argument we therefore cannot take for granted that causal effects is something that ought to be clarified by demonstrating statistical differences via RCTs. Moreover, RCTs are not necessarily the best way to clarify causal relations as they can only provide sparse information about what the relevant causal connections consist in, how they may come about, and how similar effects might be attainable in the future.

Notice that this argument is related to the arguments made by many Meaning-makers that the results of RCTs cannot tell us *why* any psychotherapy process has had this or that effect, nor *how*. However, though dispositionalists would wholeheartedly agree with the Meaning-makers that the relevant information cannot be gained by trying to establish knowledge of causal covering laws (Erklärung), pace the Meaning-makers, dispositionalism implies emphasizing an alternative account of causal explanation rather than to deny its relevance. Thus, though the Meaning-makers have put apposite emphasis on understanding agents' unique points of view (Verstehen), causality is not the enemy. Rather than to construe Verstehen as about non-causal phenomena, understanding what something means for someone is more often than not to get to know about their causally powerful dispositional properties.

In line with a Humean perspective, both the Medical model and APA's statement on evidence based psychological practice treat causal effects as something that must be clarified by RCTs. At best, this understanding is incomplete. For one thing, it ignores that there can be regular succession of events that are not causally related, and moreover, we must account for the possibility of causal processes that happen only once (the kind of result an RCT would purposely elide). To deal with these features, dispositionalism revives a realist view of causality, in which causal relations rest upon the powers of dispositional properties to produce changes. And on the view developed by Mumford and Anjum (2011), causal relations are constituted by properties that only dispose towards other properties as their effects. Causes may thus only tend towards their effects, and these effects might never be manifested in any observable regularity (Anjum and Mumford 2018a).

Thus, contra the Humean conceptions, isolating variables in the hope of measuring regular relations between them is no royal road to know about causal relations. Not only do we need more thorough inquiries that explain how and why causal effects emerge, but knowledge about relevant causal links and mechanisms can even be gained without having recorded any correlations, for instance, when the possible interplay between dispositional properties can be understood before any causal changes emerge. Statistical evidence is thus not needed if we already understand the mechanisms involved, and causal claims are best supported by theory that explains how and why causal effects are brought about (Anjum and Mumford 2018a, b). Nor is statistical evidence needed if we can come to understand how and why causal effects may emerge by reflecting on the possible interplay between the dispositional properties of persons and their surroundings. Importantly, these advancements of causal dispositionalism do not only bring support to the arguments of both Dodobirds and Meaning-makers, but as such they demonstrate the relevance of dispositionalism both for psychotherapy research and psychotherapy.

12.9 Implications for Psychotherapy Research

Of the many relevant implications of dispositionalism for psychotherapy research only five will be delved into here. That is, (i) *methodological pluralism* and (ii) *causal singularism*, and the related potentials for (iii) *advancing theoretical reflection*, (iv) *avoiding pseudo-empirical research* and advancing the (v) *theoretical integration of psychotherapy perspectives*. These interrelated aspects are picked out both because they have profound potential for changing the field of psychotherapy research for the better and because they are relevant for the implications of dispositionalism for psychotherapy discussed in the next section.

As mentioned, the statement of APA (2006) approves of a multitude of evidential resources, but from a dispositionalist perspective it is still too narrow. By upholding RCT's as the standard for drawing conclusions about the effect of psychotherapy one treats causal relations as something that must be discovered through RCTs. However, questions of how to clarify causal relations must not be conflated with ontological questions of what causality is. Thus, though RCTs can indicate the possibility of relevant causal links by demonstrating difference making and regularity on an average level, there are also other, and often better, ways to get in touch with causality. For instance, by understanding what was experienced by someone; say, when we wonder why a child suddenly got anxious when playing with his dad, and get to know that playing with toy-soldiers with his dad suddenly made the child remember that his dad could not only act unpredictably, but also just as aggressively as the toy-soldiers. Accordingly, we must emphasize *methodological pluralism* which is also related to *causal singularism*: Causality is manifested in concrete particular instances because of the dispositional properties involved (Anjum and Mumford 2018b). Whether a child becomes anxious when experiencing a parent as unpredictable, depends upon the properties of the child, the parent, and their surroundings, not on whether a statistically significant number of children experiencing their parents as unpredictable become anxious when together with their parents.

This brings us to the third implication of dispositionalism: Singularism means not only that the priority traditionally given to quantitative research rather than qualitative research is untenable, but also that we should emphasize the primacy of causal theory over statistical data. Thus, methodological pluralism does not only mean that there are many ways to get in touch with causality, but also that psychologists must put more emphasis on *theoretical reflection*. Agreeing with APA (2006: 274) that one should recognize the strengths and limitations of evidence from different types of research, we should notice that where Hume's covering law model suggests collecting more and more data of the same type from repeated instances, dispositionalism encourages us to collect more data about singular instances. What matters is that we can understand the causal processes involved in actual singular cases. What happens elsewhere and at different times is not only ontologically irrelevant but may also be epistemologically extraneous. Not only do we need more thorough inquiries that explain how and why causal effects emerge here and now, but we can also clarify possible interplay between dispositional properties before any causal changes have emerged (Anjum and Mumford 2018b).

This leads us to the fourth implication of dispositionalism, that is, the potential for *avoiding pseudo-empirical* research. The notion of pseudo-empiricism, acknowledged by most Meaning-makers (e.g. Brinkmann 2011; Valsiner 2012; Harré and Moghaddam 2012), that we should not mindlessly put assertions to empirical test if they can be evaluated by other means, was first introduced by Smedslund (1995). Notice, however, that acknowledging the relevance of this notion does not mean that we ought to follow Smedslund's arguments all the way towards concluding that psychology cannot be an empirical science. On the contrary, not only is it probably more coherent to think of theoretical reflection as comprising kinds of inquiries in which experience plays a significant role (cf. Casullo and Thurow 2013), but in line with methodological pluralism, RCTs and other kinds of empirical research do have their areas of application. Thus, acknowledging the relevance of avoiding pseudo-empirical research only means that an emphasis on empirical research must not exclude the relevance of theoretical reflection and that the relevance of advancing theories of causal tendencies is often more relevant than statistical data.

For instance, collecting more empirical data to test the hypothesis that people tend to attempt to do what they think they are able to do when they also want it, or whether people tend to get anxious when together with unpredictable people, will be pseudo-empirical. Thus, if we already have relevant and sufficient amount of information about the relevant dispositional properties, we can know this by reflecting on the possible interplay between them without further empirical inquiry. For instance, if we know that someone (say, Caesar) has come to think that he is vulnerable, and if we also know that he thinks that someone else (Brutus) may come to do something (stab with a knife) that might kill or hurt him, and he also thinks that he cannot know whether this may be done at a time when he is able to prevent it, then we know that he is disposed to become anxious when together with Brutus. Moreover, we can also know by reflecting on possible preventing circumstances, that these are only tendencies. For instance, though Brutus thinks that he is able to stab Caesar and wants it, he might not attempt to do it because he thinks that Caesar's guards may be able to protect Caesar, and Caesar may feel safe with Brutus when his guards are around. Similarly, there are no unpreventable laws to be found that children having unpredictable parents will become anxious when together with their parents. They may feel safe in context of their grandparents, by believing they are stronger and/or more competent than their parents, etc. Nevertheless, knowing about such tendencies is practically relevant, and we should avoid being unpredictable with regards to potentially harmful actions if we want to become worthy of anyone's trust, our children included.

Though these examples may sound trivial, the point itself is not. Though it is a matter of controversy how much psychological research has been pseudo-empirical, a growing number of psychologists have recognized the relevance of the notion. Indeed, the point does not only extend to many psychological theories, many of which are clinically relevant, but also to psychotherapy research. A striking example is the aforementioned Dodo-bird-claim that the great mass of empirical data actually produces evidence that falsifies the Medical model. There is nothing wrong with this conclusion, except that it is *pseudo-empirical.* In other words, that the practical relevance of psychotherapy perspectives must be tested on an average level by RCTs – as if psychotherapy was some kind of context-transcending pill with regular and replicable effects working independently of the unique properties of the persons involved – is simply nonsense. This is adeptly demonstrated by the above-mentioned Meaning-maker-argument that RCTs cannot ever pay due respect to the fact that existing persons are someone for whom something exists: As none of us will ever make sense of things from the exact same perspective as any other, and no experience can ever be undone, and moreover, that we are continuously open for change by attaching new meanings to things, implies that we cannot take for granted that persons will react in the same or similar ways to neither the same nor similar events. Hence, there is no other option than to qualify our services one psychotherapy process at a time.

This also brings us to the fourth potential of dispositionalism for psychotherapy research mentioned above. Alas, the prevailing emphasis on RCTs has thrown a plethora of psychotherapy perspectives, models and theories[1] into pointless rivalry, needlessly competing for the best results on an average level. This has been at the expense of theoretical work to clarify the extent to which these perspectives may be integrated. However, the idea of psychotherapy perspectives as consisting of competing empirical hypotheses of regular causal relations between isolated variables is misleading, and the many various perspectives, models and theories are better characterized as compatible and/or overlapping attempts to put possible relations between the clinically relevant dispositional properties of persons into words. This also makes it relevant to consider the extent to which the perspectives can and ought to be theoretically integrated. The further upshot is that such integrative work will

---

[1] A far from complete list would include Narrative Therapy, Cognitive Behavioural Therapy, Schema Therapy, Meta-cognitive Therapy, Acceptance and Commitment therapy, Emotion-focused Therapy, Gestalt Therapy, Client- and Person Centred Therapy (not to be confused with person-centred medicine), Compassion-focused Therapy, Existential- and Humanistic Therapy, Mindfulness, Rational Emotive Therapy, as well as various psychoanalytically oriented and psychodynamic perspectives such as Intensive Short Term Psychodynamic Therapy, Self-Psychology, Mentalization Based Therapy, Object Relations Therapy, Traditional Psychoanalysis, Relational Psychodynamic Therapy, and many more.

highlight the relevance of a capacity that is not only pivotal for psychological research, but also vital for any psychotherapy process: To study and take part in the ever-evolving unique and vastly complex contexts of psychotherapy processes requires that we take advantage of our capacity for thorough theoretical reflection, so as to critically calibrate our knowledge of possible and impossible relations between the dispositional properties that persons may have.

Notice that philosophers sometimes speak of dispositional properties by using the shorter term "dispositions". However, to avoid confusion it should be noticed that psychologists have often used the latter term in another sense, that is, to refer to character traits by which one can purportedly predict behaviour by referring to the frequency of past behaviour. For instance, one may think of one's father as dangerous now or in the future simply because he has often done hurtful things in the past. By contrast, dispositionalism emphasizes a notion of dispositions (causal powers) as constituted by intrinsic properties, or propensities (see Rocca, Chap. 3, this book). Thus, whether a father is dangerous or trustworthy here and now, depends upon his properties here and now not on how often he has had these properties before. For instance, whether he comes to do something that hurts his child or not does not depend upon whether he has been violent in the past, but on whether he actually cares for and understands his child, on whether he currently has the relevant amount of self-control, on whether he is currently able to act autonomously, etc. Accordingly, developing relevant clinical competency depends not so much on being able to attribute statistically based character traits, as on being able to calibrate one's notions of possible and impossible relations between the dispositional properties that persons may have in various singular cases here and now.

12.10 Implications for Psychotherapy

Thus, where the Humean commitments of the Medical model lead to a narrow emphasis on empirical research and a one-sided compliance to empirically supported treatments, dispositionalism implies pluralism with respect to both research methodology and practice. In significant respects, this pluralist stance of dispositionalism is related to the proposals of a so-called *bricoleur-model* of psychotherapy (Smedslund 2012b). On this view, practitioners should not let an emphasis on specific pre-construed models and interventions stand in the way for focusing on the needs of the unique patient, but should rather be prepared to use *whatever is at hand* that might contribute to solve the problems encountered.

Alas, the need for emphasizing a practice where one strives to meet persons as openly and unprejudiced as possible so as to ensure a sufficiently flexible adjustment to the unique case, is not only different from the Medical model, but it has been disregarded by governmental authorities aiming to ensure the quality of the therapy process from the outside. The implementation of the so-called "Quick Psychological Health Services" in Norway (Norwegian: Rask psykisk helsehjelp) inspired by the "Improving Access to Psychological Therapies Programme" in

England, may serve to illustrate this. The primary aim of this programme, to provide people with free, low threshold, professional aid when suffering from depression and anxiety, is impeccable and I'm proud to have contributed to fulfilling this aim for 5 years. However, the programme has also suffered from one great mistake dictated by the dogmas of the Medical model, that is, it has exclusively emphasized cognitive behavioural therapy (CBT). Presumably, this is because it has been maintained that it has been demonstrated by RCTs that CBT *is* effective for relieving depression and anxiety. In line with the Medical model this understanding accounts for CBT as an empirically supported treatment, that is, a kind of "miracle drug" comprising statistically proven effective specific interventions. However, though the principles of the CBT-models undoubtedly contribute to understanding psychological phenomena in ways that *may* be practically relevant, this picture is seriously flawed. First, CBT is not the only perspective providing relevant assertions, neither for low threshold services, nor for dealing with depression in general. Second, it may not always be the most relevant one, and third, if it is relevant, it may be so in combination with other perspectives.

However, though cognitive-behavioural-therapists have often been eager to promote CBT as an empirically supported treatment they have also often described matters as being more complicated than as construed by the Medical model. Rather than picturing persons attending psychotherapy as patients who passively receive some kind of miracle drug, one has not only emphasized the relevance of establishing a trusting relationship, but also of motivating clients to contribute as active agents towards reaching their goals. However, how various dispositional properties of patients, therapists and their surroundings may combine and intertwine towards establishing a trusting relationship are not described by CBT-models, but more extensively by other psychotherapy-perspectives (e.g. various psychoanalytically oriented and psychodynamic perspectives). Also, though one actively encourages patients to take active part in the process, this is normally done by "socializing the patients" towards proceeding within the frame of specific CBT-models, and this way of proceeding puts us right back into the linear scheme of the Medical model; as if all patients always suffer in ways to be treated by the same treatment procedures.

Yet, that one size does not fit all, is also well known among cognitive-behavioural therapists. E.g. it has been recognized that it is not always sufficient to deal with the *content of cognitions* triggered in specific situations to counteract various problems. For instance, realizing that your first thought that someone did not like you actually was wrong – and that it was more likely that the reason this person avoided you was that he was shy – may not be enough to overcome social anxiety. Rather, or additionally, more *fundamental and context-transcending assertions* (often called schemas) may have to be dealt with. For instance, being convinced that no one will ever really like you if you fail at something, and thus, that you have to make sure not to fail in any kind of situation. To take such more fundamental assertions into account so-called Schema-focused models of psychotherapy was construed by integrating aspects from CBT and psychodynamic perspectives. More recently, the fact that it is not always sufficient, sometimes not necessary, and perhaps not even desirable, to alter the content of cognitions has been incorporated into the therapy-models. For instance, we may not want to change our minds about it being both true and sad that someone we love is no longer alive. Thus, one has acknowledged the relevance of working with how to deal with having thoughts without altering their content by integrating Mindfulness exercises, for instance, practicing on recognizing one's thoughts as a natural and harmless aspects of the fluctuating moment, or by bringing in the principles of Meta-cognitive therapy, for instance, by recognizing that scary thoughts about something are actually not dangerous in themselves.

From the perspective of dispositionalism, this picture could not only easily, but also ought to, be expanded by aspects of further therapy models and perspectives not traditionally thought of as part of the cognitive paradigm. For instance, one may take into account the aspect emphasized by models of Emotion-focused therapies that simply work on how to think differently without making sure that this actually creates emotional change, may not be sufficient, or one may aptly integrate aspects traditionally emphasized by various psychodynamic perspectives such as Object-relations therapy and Self-psychology. For instance, validating the experiences of patients (making their thoughts understandable without necessarily agreeing with their content) so as to foster self-esteem by the recognition of the value and intelligibility of being who they are (this may have been prevented by inhospitable circumstances when growing up, insensitive or violent parents and/or by bullying at school). Or one may inquire into whether inhospitable growing up conditions have made persons prone to satisfying other people's needs rather than their own, perhaps even to the extent of having become unable to recognize their own real wants and needs, eventually leading to depression or anger. Such inquiries may have the aim of providing the patients with the relevant understanding of themselves both to prevent feeling ashamed about one's suffering and to foster opportunities for exploring how to flourish in one's life on one's own premises.

However, as aptly pointed out by Wampold (2019) there is a risk that the idea of integrating various approaches will be taken to be a specific approach in its own right, so that one simply adds to the expanding number of specific therapies that purportedly must be evaluated by RCTs. Fortunately, dispositionalism offers a viable solution to overcome this risk. Thus, the aim of clarifying the extent to which various perspectives are overlapping or may be fruitfully combined is not to construct yet another empirically supported treatment, but to get a hold on how the various perspectives describe possible causal links between the possible dispositions of persons that *might* be relevant in possible singular and unique cases. The aim of integration is not to construct yet another miracle drug suitable for all, but to widen one's scope, so as to be more flexible and able to deal with the immensely complex ways in which the various dispositional properties of unique persons may interact both with each other and the properties of the surroundings. As such, theoretical integration is no aim in itself, but may serve the more general bricoleurious aim of being prepared to use *whatever is at hand* that might contribute to solve the problems encountered. Thus, this overall aim does not only include taking various psychotherapy perspectives into account, but also perspectives and ideas suggested by professionals primarily working with other kinds of services than psychotherapy, such as social workers and physiotherapists, as well as anything else that might be beneficial. For instance, making a phone call to a social worker to get help solving financial problems, may be no less relieving than psychotherapy. Or by taking into account how moods may change by altering body-posture, helping patients to recognize that though their traumatic experiences may have made them more prone to strain their muscles (e.g., as part of being angry or terrified), become restless or agitated (e.g. as part of being worried or more alert) or collapsed (e.g. as part of being depressed or shameful) as part of a bodily defence against perceived threat, their inborn ability to create a more balanced and harmonious posture is not destroyed.

Thus, where the Medical model portrays clinical competency as the ability to use psychological knowledge in such a way that it leads to statistically probable change, dispositionalism emphasises our abilities to critically calibrate our knowledge of the vast amount of possible and impossible relations between the dispositional properties that persons may have in various circumstances. In other words, though we should not deny ourselves the possibility of looking to RCTs for inspiration, making it obligatory to look to this kind of research for evidence runs the risk of distracting focus away from what matters, which is to deal with the properties and processes involved in actual singular cases. Just as one does not necessarily get wiser simply from having had more experience, we cannot know whether relying on RCTs conducted elsewhere and with other patients at another time is relevant here and now. Accordingly, just as clinical competency cannot be assured through inductions from the unavoidably limited and biased experience of individual therapists, neither can it be built on inductive generalisations from accumulations of empirical data from RCTs or other correlational studies alone. Rather, the clinical competency relevant for psychotherapy has to do with having gained the relevant *degrees of interpretational freedom* for dealing with possible tendencies. The extent to which humans really differ from other animals with respect to the capability for building such competency might be discussed. What is clear however, is that human beings are normally able to recognize more opportunities, say, with a nut than eating it raw (e.g. baking a cake). Similarly, we not only can but should be more flexible than simply complying with empirically supported treatments. By working to clarify the possible and impossible relations between the dispositional properties that persons may have in various contexts, one might not only get a reflective overview of various possibilities that may be actualized in concrete situations, but one might also gain relevant resources for avoiding overgeneralizations and avoiding jumping to unwarranted conclusions. Thus, clarifying possible links between the dispositions of persons may not only strengthen an apposite sensitivity for more possibilities than one's immediate first impressions, but it might also provide resources for understanding and dealing with actualized feelings, thoughts and behaviour.

12.11 As Statistics Don't Get It, Try Getting the Vectors Right

To the extent that the above discussions are up to something, dispositionalism does not only improve earlier critique of the Medical model, but may as such also have profound consequences for psychotherapy research and mental healthcare services. Dispositionalism provides an account of causality that implies pluralism with respect to both research methododology and practice, all toward the aim of dealing sufficiently with the complex interplay between the unique properties of persons, their circumstances, and clinical practitioners. This can be illustrated by the vector model suggested by Mumford and Anjum (2011) and introduced in Chap. 2 in this book.

Thus, where the Medical model portrays all psychotherapists as complying with the same empirically supported treatments purportedly making a statistically probable difference on specific symptoms of specific disorders purportedly shared by various patients (see Fig. 12.1 above), the vector model (see Figs. 12.2 and 12.3) provides a way to emphasize and clarify aspects of a vastly more complex reality consisting of both catalysing and preventing dispositional properties of patients, therapists, and their circumstances.

The broken line T on top in the picture represents a threshold that has to be met for the experience of improvement and/or wellbeing to occur for some patient P, while the broken line at the bottom of the picture represents some threshold condition for the experience of suffering or of having some kind of problem. The horizontal lines P and S represent some points in time where the various properties of the patient and various conditions of his/her current situation, respectively, push and

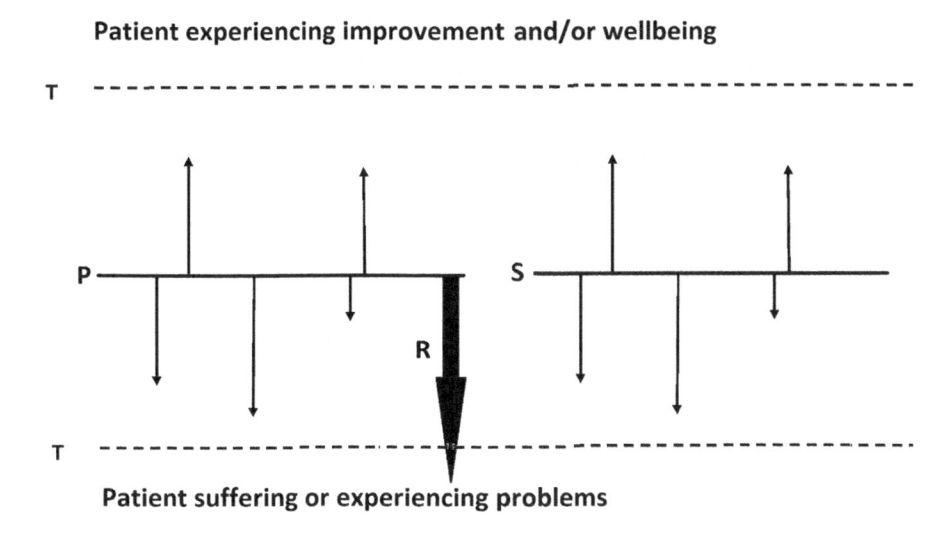

Fig. 12.2 The vector model part 1: patient and his/her situational circumstances

Patient experiencing improvement and/or wellbeing

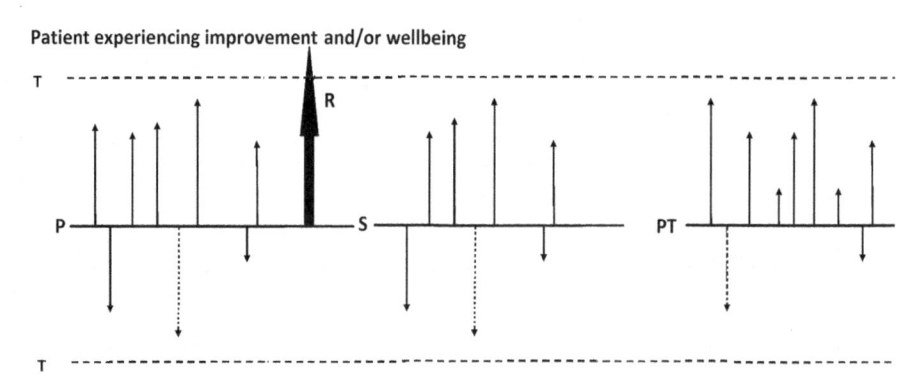

Patient suffering or experiencing problems

Fig. 12.3 The vector model part 1: patient, situational circumstances and psychotherapist

pull towards the thresholds for experiencing wellbeing or suffering. The arrows pointing upwards from the horizontal line P represent the properties of the patient that dispose towards wellbeing, while the arrows pointing downwards represent the properties of the patient that dispose towards suffering. The thicker arrow R represents the overall result of all the vector-dispositions; alas, at this point in time, crossing the line for suffering to emerge. Hopefully, however, there are some potentials for helping the patient. First, there are already some dispositions still pointing upwards, and if it could be possible either to strengthen these or to add some vectors pointing in the upwards direction, the overall resultant vector R might turn in the other direction, and hopefully dispose as far as reaching not only improvement, but also well-being. Alternatively, or additionally, the vectors pointing downwards could be removed or at least weakened, so as to bring the overall result less in the direction of suffering and more in the direction of wellbeing. However, at this point in time, the properties of the patient and his/her surroundings leave the patient stuck in suffering.

Say, however, that one of the properties of the situation involves a friend who encourages the patient to consult a psychotherapist, and even better, one of the properties of the patient is that he/she thinks of the friend as trustworthy. Fig. 12.3 illustrates what could ideally have happened if the patient and the psychotherapist met. Now that the psychotherapist has become part of the complex situation, the horizontal line PT represents the period in the patient's life when consulting the therapist, and the arrows pointing upwards from PT represent properties of the psychotherapist that dispose towards the wellbeing of the patient. The arrows pointing downwards, however, represent the properties of the psychotherapist that dispose towards worsening the suffering of the patient. For instance, the therapist may be prone to misunderstand, or act in ways that the patient experiences as too challenging.

Fortunately, however, though there are arrows pointing downwards, the downward arrow in the far right of the figure is rather weak compared to the other vectors, and moreover, the vectors pointing upwards represent properties of the therapist that dispose towards helping the patient. For instance, the therapist cares for the patient, dares to call a social worker for relevant financial advice or aid, knows about the various ways that actions, thoughts, and emotions may relate to various experiences, does not let statistical considerations take focus away from the actual patient, has sufficient amount of self-control, and so on.

Moreover, the other arrow pointing downwards from PT is dotted, which represents that the therapist has had the ability to prevent some property of him- or herself from contributing negatively to the therapy process, for instance, his disagreement with the patient's political views, being hungry, or prone to think about his/her own family-problems. Notice also, that during the therapy process captured by Fig. 12.3, the number of patient- and situation- vectors pointing upwards has increased compared to Fig. 12.2, and although this could possibly have happened independently of the psychotherapy process, it might also be the result of, or perhaps even dependent upon, the therapy-process. Moreover, one of the vectors pointing downwards from P, as well as one from S, are now dotted. Again, this represents that they are removed from the complete set of properties relevant for the mental health of the patient. For instance, in line with the examples described in the former section, some financial problems may be gone, weakened or prevented by making contact with a helpful social worker. Or the patient has become able to recognize that he/she has tended to take more care of others than him/herself, and has additionally managed to forgive him/herself for this by having worked with the therapist to understand these actions as having had an understandable rationale. Moreover, this may have contributed to release potential for working towards being able to act more autonomously, flexibly and in tune with one's own desires. Thus, the overall result R is now different from in Fig. 12.2, pointing upwards.

There is clearly much more work to do in order to clarify the implications of dispositionalism for psychotherapy research and mental health care. However, the take-home message so far is that nothing (except from ethical concerns) should stop us from doing *whatever it takes* to find out about and to deal with the relevant dispositional properties involved. Statistics may help to indicate the existence of relevant causes in larger groups and on an average level. And if we are lucky, these causal tendencies are relevantly similar to what happens when encountering unique individuals here and now. However, beyond that, statistics don't help us much. Thus, we are rather in need of thicker explanations, a deeper understanding of the properties and experiences of persons and their contexts that dispose towards well-being and suffering. If statistics don't get it, try getting the vectors right.

References and Further Readings

American Psychological Association Task Force on Evidence Based Practice (2006) Evidence-based practice in psychology. Am Psychol 61:271–285

Anjum RL (2016) Evidence based or person centred? An ontological debate. J Eval Clin Pract 4:421–429

Anjum RL, Mumford S (2018a) What tends to be; the philosophy of dispositional modality. Routledge, London

Anjum RL, Mumford S (2018b) Causation in science and the methods of scientific discovery. Oxford University Press, Oxford

Brinkmann S (2011) Towards an expansive hybrid psychology: integrating theories of the mediated mind. Integr Psychol Behav Sci 45:1–20

Casullo A, Thurow J (eds) (2013) The a piori in philosophy. Oxford University Press, Oxford

Chambless DL, Hollon SD (1998) Defining empirically supported therapies. J Consult Clin Psychol 66:7–18

Crits-Christoph P, Chambless DL, Markell HM (2014) Moving evidence-based practice forward successfully: commentary on Laska, Gurman, and Wampold. Psychotherapy 51:491–495

Duncan BL, Miller SD, Wampold BE, Hubble MA (2010) The heart & soul of change: delivering what works in therapy, 2nd edn. American Psychological Association, Washington, DC

Groff R (ed) (2008) Revitalizing causality: realism about causality in philosophy and social science. Routledge, London

Groff R (2011) Getting past Hume in the philosophy of the social science. In: Illary PM, Russo F, Williamson J (eds) Causality in the social sciences. Oxford University Press, Oxford, pp 296–316

Groff R, Greco J (eds) (2013) Powers and capacities in philosophy; the new aristotelianism. Routledge, London

Harré R (1999) Rediscovery of the human mind. Asian J Soc Psychol 2:43–62

Harré R (2002) Cognitive science. A philosophical introduction. Sage, London

Harré R, Gillett G (1994) The discursive mind. Sage, London

Harré R, Madden EH (1975) Causal powers: a theory of natural necessity. Blackwell, Oxford

Harré R, Moghaddam F (eds) (2012) Psychology for the third Millenium: integrating cultural and neuroscience perspectives. Sage, London

Hollon SD (2006) Randomized controlled trials. In: Norcross JC, Beutler L, Levant RF (eds) Evidence-based practices in mental health; debate and dialogue on the fundamental questions. American Psychological Association, Washington, DC, pp 96–105

Kennair LEO, Aarre TF, Kennair TW, Bugge P (2002) Evidence-based mental health – the scientific foundation of clinical psychology and psychiatry Scipolicy™. J Sci Health Policy 2

Mumford S, Anjum RL (2011) Getting causes from powers. Oxford University Press, Oxford

Orlinsky DE, Rønnestad MH, Willutzki U (2004) Fifty years of psychotherapy process-outcome research: continuity and change. In: Lambert MJ (ed) Bergin and Garfield's handbook of psychotherapy and behavior change. Wiley, New York, pp 307–389

Smedslund J (1995) Psychologic: commonsense and the pseudoempirical. In: Smith J, Harré R, Van Langenhove L (eds) Rethinking psychology. Sage, London, pp 196–206

Smedslund J (2009) The mismatch between current research methods and the nature of psychological phenomena: what researchers must learn from practitioners. Theory Psychol 19:778–794

Smedslund J (2012a) Psycho-logic: some thoughts and after-thoughts. Scand J Psychol 55:295–302

Smedslund J (2012b) The bricoleur model of psychotherapeutic practice. Theory Psychol 22:643–657

Stiles WB (2009) Responsiveness as an obstacle for psychotherapy outcome research: it's worse than you think. Clin Psychol 16:86–91

Teigen KH (2002) One hundred years of laws in psychology. Am J Psychol 115:103–118

Valsiner J (2012) A guided science: history of psychology in the mirror of its making. Transaction Publishers, New Brunswick

Valsiner J (2014) Breaking the arrows of causality; the idea of catalysis in the making. In: Cabell KR, Valsiner J (eds) The catalyzing mind; beyond models of causality. Springer, New York, pp 17–32

Valsiner J (2017) From methodology to methods in human psychology. Springer, New York

Valsiner J, Brinkmann S (2016) Beyond the "variables": developing metalanguage for psychology. In: Klempe SH, Smith R (eds) Centrality of history for theory construction in psychology. Springer, Cham, pp 75–90

Wampold BE (2019) The basics of psychotherapy; an introduction to theory and practice. American Psychological Association, Washington, DC

Wampold BE, Imel ZE (2015) The great psychotherapy debate; the evidence for what makes psychotherapy work. Routledge, London

Open Access This chapter is licensed under the terms of the Creative Commons Attribution 4.0 International License (http://creativecommons.org/licenses/by/4.0/), which permits use, sharing, adaptation, distribution and reproduction in any medium or format, as long as you give appropriate credit to the original author(s) and the source, provide a link to the Creative Commons license and indicate if changes were made.

The images or other third party material in this chapter are included in the chapter's Creative Commons license, unless indicated otherwise in a credit line to the material. If material is not included in the chapter's Creative Commons license and your intended use is not permitted by statutory regulation or exceeds the permitted use, you will need to obtain permission directly from the copyright holder.

Chapter 13
Causal Dispositionalism and Evidence Based Healthcare

Roger Kerry

13.1 Complexity in Practice

When I was a young Junior Physiotherapist, a department I worked for was chosen as a centre to collect data for one of the first multi-centred low back pain trials in the UK. My job was to identify any of my low back pain patients to see if they could be included in the trial. I was keen to be involved, as I was very curious about what might work for people with low back pain. People with low back pain typically struggle to respond well to many interventions, hence the global epidemic status of non-specific low back pain. I was surprised (naively) that the trial criteria excluded people with co-morbidities, poor general health status, and repeated previous treatments – this would be most of my patients. When I had the chance to discuss this with the visiting trial co-investigator, he explained that they were trying to get the conditions of the trial as controlled as possible so that they could get a "good, clear view" of what worked for people with low back pain. The identification of causal associations between two variables is of course the scientific ideal. The removal and control of possible confounders is a hallmark of high-quality health research. Yet this is not what my patients looked like. What was the impact of their co-morbidities? How did their general health affect their painful experience? What were the biological and psychological consequences of repeated previous treatment? It was clear that the trial conditions were not representative of the complex and context-sensitive clinical shop-floor.

My interest in the CauseHealth project stems primarily from questions raised on the shop-floor of clinical healthcare – and specifically physiotherapy. As a clinician, teacher, and researcher in physiotherapy, there has been one question in particular, which I first began thinking about many years ago during the emergence of evidence based healthcare in the early 1990s when I was training as a physiotherapist.

R. Kerry (✉)
Faculty of Medicine and Health Sciences, University of Nottingham, Nottingham, UK
e-mail: Roger.kerry@nottingham.ac.uk

Is the sort of causation we establish in population research studies the same sort of causation we are seeking with an individual patient?

Witnessing contradictory results on the shop-floor – for example, a patient not responding to a strongly 'evidence based' intervention – is not uncommon and can easily be passed off by considering that person as a "non-responder". However, is this always the case? Or is there something about the individual context which is not readily represented in population studies? If so, what are the fundamental assumptions about causation in both clinical practice and scientific, systematic research? With all this in mind, my philosophical interest in causation began, and my relationship with CauseHealth was soon cemented.

Causation lies at the heart of healthcare. This is clearly manifest in aspects of healthcare such as causes of disease, but of course it is also the core focus of research methods which try to establish the therapeutic effectiveness of health interventions: what causes someone to get better. Such research methods represent a central feature of the practice framework of evidence based healthcare (EBHC) – that is the proposal that all healthcare practice and decision making should be based on the *best available evidence*. This evidence is multi-factorial and can in principle emerge from a number of sources, such as systematic clinical research, laboratory research, clinical observations, experience, and so on. The most recent formal iterations of EBHC stems from literature regarding evidence based medicine in the early 1990s (Guyatt 1991; Guyatt et al. 1992; Sackett et al. 1996). This literature sets out specific and detailed structures by which different evidential sources are given different priorities and is still explicitly reflected in the most contemporary re-workings of EBHC, for example the GRADE framework (Mercuri and Baigrie 2018; Guyatt et al. 2011). Philosophically, this is a critical feature of EBHC, as it allows us to understand some fundamental assumptions on which clinical decision making is based. If certain evidential sources are given priority over others, we have an insight into the epistemological and ontological groundings of what we do as healthcare professionals, in both practice and research.

In this chapter I therefore summarise an early and central idea from the CauseHealth project, which served to motivate much of the direction of thought and research within the project. That idea was that the way in which EBHC prioritises its evidential sources demonstrates that the underpinning causal theory of healthcare is essentially Humean (Kerry et al. 2012). To problematise this, we sought to consider what the limitations of practicing healthcare were if a strict Humean notion of causation was assumed. In doing so, it became clear that perhaps what the evidential structuring of EBHC meant by causation was not necessarily the sort of causation which is at play in real-life, complex healthcare decision making. As such, we offered an alternative causal theory, based on dispositionalism. This theory could better account for both the scientific processes (methods) used to establish data from populations, and the context-sensitive, complex mass of information witnessed in individual clinical encounters. Where Hume saw causes as nothing more than regularly occurring events with certain temporal and spatial relationships, while Hume's opponents argue that, in addition, causes necessitate their effect, we take causes as real features of the world which interact with each other and have only a tendency to manifest in an effect (Kerry et al. 2012). Dispositionalism, I will argue, offers a more real-world account of causation.

13.2 Evidential Hierarchies Expose Causal Theory

During my formative years as a physiotherapist, I was schooled in the ways of best practice and the limitations created by our *human biases*. I would often be the first to dismiss any enthusiasm which other 'lesser' colleagues might experience when patients responded well to our interventions: for example, a patient claiming they had a huge reduction in pain after some hands-on therapy. My responses would always be along the lines of "you don't know that it was the treatment that made the difference", or "there is no RCT-level evidence to support that", and "you have to be aware of your own perception biases", and so on.

But what if that patient had been a subject as part of a trial? What if the trial protocol was just the same as what had happened in clinic that day? The improvement seen would then be recorded and added to data from other similar subjects, and then compared to data from subjects who had a different intervention, or whatever. That response would then be considered positive and causally associated to the intervention. But this causal claim would not be made on what had been observed that day, but rather what had happened to other subjects.

This counterfactual reasoning is again part of the essence of our research structures. However, dispositionalism would suggest that the causal process was always a real feature of what happened in that individual case. Can the clinician be their own scientific data collector, ensuring critical analysis, reflection, and a systematic approach to patient care, and in doing so observe real causation? A dispositionalists account of causation would permit this. The reaction from EBHC is that you cannot make generalised claims from such individual observations. We would agree. Dispositionalism takes all causal cases as context-sensitive, so what works in one context is not what would work in another one.

I will now demonstrate how a Humean notion of causation can be read from evidential hierarchies. I explain some of the problems with this and present four key *desideratum* to highlight how the dispositionalist alternative can address the shortfalls of Hume's approach to causation.

If we take the inaugural papers of Guyatt, Sackett et al. (as above) as the onset of the contemporary working of evidence based healthcare (originally referred to as evidence based medicine), the structuring of evidential sources is clear. Essentially, research methods with a low risk of bias for confounding are given evidential priority. Randomised controlled trials (RCTs) are considered a gold standard method as they are able to control for both known and unknown confounding variables. In doing so they, ideally, isolate the hypothesised causal variables from other possible causes and observe the effect in repeated conditions (large samples, repeated studies, etc.). The data from such trials is then sought, to be synthesised and presented in systematic reviews, which sit at the top of accepted evidential hierarchies. Methods with increasing risk of bias are progressively de-emphasised down the hierarchy. This is in line with what La Caze (2008) refers to as a *categorical* reading of evidential hierarchies. High quality RCTs and low risk observational studies thus provide confidence in causal claims, whereas de-emphasised sources do not. This structuring is honourable and adheres to fundamental principles of science. It also gives a direct indication of the causal assumptions which healthcare works on.

Consideration of the characteristic differences between 'causal methods' (RCTs etc.) and 'non-causal methods' (case studies, mechanistic studies, experience, etc.) provides a starting point here. What defines these causal methods is the fact that they first systematically observe the relationship between two events (say, A and B) multiple times whilst controlling for possible confounding, and then, facilitated by statistical modelling based on a frequentist idea of probability (see Rocca, ch. 3, this book), make a judgment on whether causation exists, i.e. whether the intervention (A) causes a desired health improvement (B). The difference between RCTs and observational studies is that the observations within RCTs are then compared to observations in another group which does not include the variable of interest (i.e. A). The de-emphasised methods and evidential sources do not possess these features.

With these characteristics in mind, it is now possible to analyse these processes a little further in order to draw out what ontological background these methods, and the associated causal claims, operate on.

Even with a superficial reading, it is clear to see a Humean essence to the causal ontology: Hume allowed that causation could be wholly represented in fact by adherence to three criteria; temporal priority, contiguity, and constant conjunction:

> ... we may define a cause to be an object, followed by another, and where all the objects similar to the first are followed by objects similar to the second. (Hume 1748 EUH 7.1.60)

Further,

> Every object like the cause, produces always some object like the effect. Beyond these three circumstances of contiguity, priority, and constant conjunction, I can discover nothing in the cause. (Hume 1740: A 9)

So, the cause always precedes the effect (A precedes B in time), the effect is consistently close to the cause (A and B are spatiotemporally contiguous), and the association is repeatedly and constantly observed (events like A are invariably followed by events like B). We can thus claim causation in a Humean sense (A causes B). This is a *regularities* view of causation, which is typically Humean and satisfies many tenets of conventional scientific principles.

This analysis can now take a short step further if RCTs are considered separately. If we assume that the characteristic difference between RCTs and observational studies is *in essence* the presence of a comparative group (current treatment, control, placebo), then there is a further aspect of the causal theory which can be considered in a Humean sense. This is the impact of *counterfactual dependency*, and also identifying the *truthmaker* of causation.

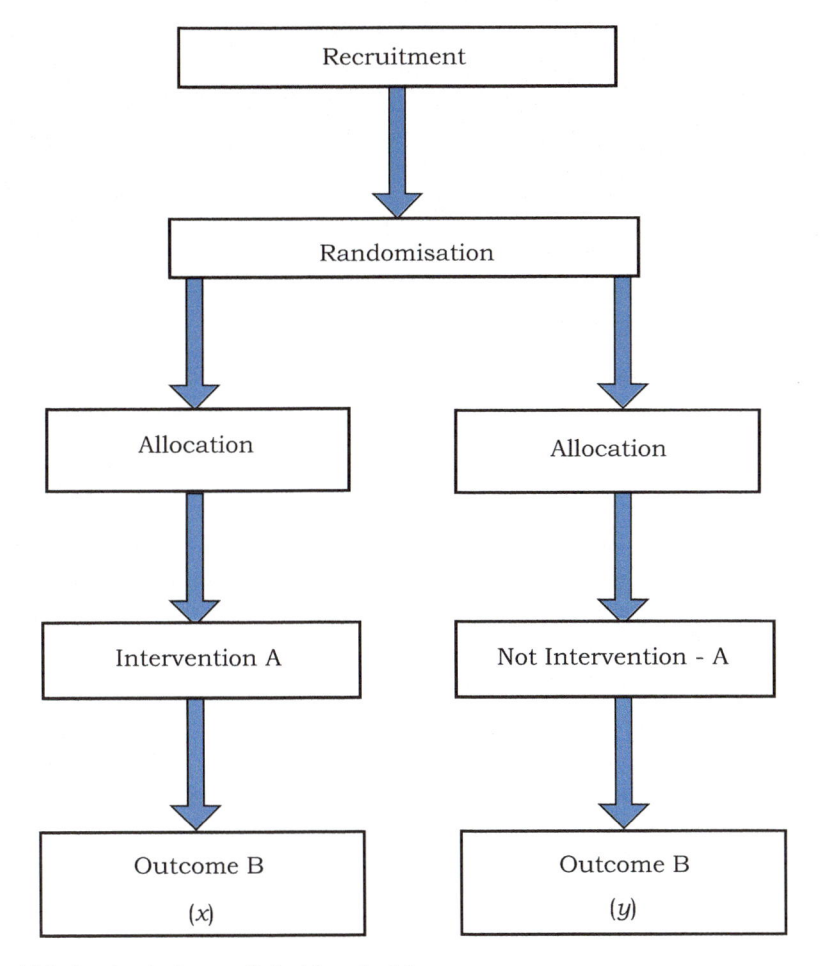

13 Causal Dispositionalism and Evidence Based Healthcare

```
                    ┌─────────────────┐
                    │   Recruitment   │
                    └────────┬────────┘
                             ↓
                    ┌─────────────────┐
                    │  Randomisation  │
                    └────────┬────────┘
                     ┌───────┴───────┐
                     ↓               ↓
             ┌──────────────┐ ┌──────────────┐
             │  Allocation  │ │  Allocation  │
             └──────┬───────┘ └──────┬───────┘
                    ↓                ↓
             ┌──────────────┐ ┌──────────────────┐
             │ Intervention A│ │ Not Intervention - A│
             └──────┬───────┘ └──────┬───────────┘
                    ↓                ↓
             ┌──────────────┐ ┌──────────────┐
             │  Outcome B   │ │  Outcome B   │
             │     (x)      │ │     (y)      │
             └──────────────┘ └──────────────┘
```

Fig. 13.1 Randomised controlled trial methodology
The randomisation and allocation processes create the counterfactual conditions by which $B(x) - B(y)$ would, under a counterfactual account of causation, constitute a causal claim. However, causation is happening in each group, irrespective of the other group.

The fundamental set-up of RCTs is represented in Fig. 13.1. This illustrates a typical, simple comparative study. The left-hand group is the intervention group (A). The right hand is the counterfactual (¬A), which is the control in which not-A. In the intervention group a level of outcome is observed, $B(x)$. In the counterfactual control group, outcome $B(y)$ is obtained. The Randomisation and Allocation stages provide the strict counterfactual conditions. If the truthmaker was the counterfactual, then causation is $B(x) - B(y)$. This satisfies that A➔B. Without y, and therefore without the control ¬A, there could be no claim to causation.

So, a comparative group can be thought of as a *counterfactual* condition. This is a condition wherein the variable of interest (the therapeutic intervention under investigation) does not exist but where as much as possible otherwise remains the same. Neo-Humean philosophers refer to this as the *closest possible world* (Lewis 1973). In RCTs, the counterfactual acts as the *truthmaker* to causation. Here's how: let's say that in Group 1 (intervention), 62% of all participants achieved a positive outcome. If this were an observational, uncontrolled study, all we could say is that there is some sort of correlation, to the magnitude of 62%, between the variable *A* and variable *B*. We would not be confident in drawing causal claims from that because other variables (confounders) could in fact be the cause of *B*. So, another group is set up which does not include *A* (the counterfactual), and if there was a similar response rate, we might logically say that the cause of the 62% positive responses was not because of *A*. Alternatively, if there was a much lower response rate in the counterfactual group, for instance 30%, we might (using statistical methods) say that the difference is sufficiently significant such that we are confident that *A* was the cause of *B*, because not half as many participants in Group 2 achieved a positive response to the intervention.

In either of these scenarios, the element of the methodological make-up which gave us confidence in inferring causation from Group 1 was not in fact what happened in Group 1, but rather what happened in Group 2, the counterfactual. It wasn't until we observed the response rate in the counterfactual group that we were confident to read causation into what happened in Group 1; thus Group 2, the counterfactual, is taken as the real *truthmaker* of causation. This is in line with the Humean conception of causation, which also has this counterfactual element:

Or in other words where, if the first object had not been, the second never had existed. (Hume 1748 EUH 7.1.60)

We now have two distinctly Humean aspects of causation identified within the hierarchies of evidence for EBHC: a *regularities view*, and *counterfactual dependency*.

To complete the Humean picture, note that because causation is drawn from regularly occurring patterns of events and/or counterfactual conditions, any allusion to an actual *substance* or *matter* of causation is absent. That is, the processes or mechanisms of how and why *A* causes *B* is missing, or at least considered unnecessary on this causal account. Indeed, Hume said as much:

Every object like the cause, produces always some object like the effect. Beyond these three circumstances of contiguity, priority, and constant conjunction, I can discover nothing in the cause. (Hume 1740: A 9)

and,

The impulse of one billiard-ball is attended with motion in the second. This is the whole that appears to the outward senses. The mind feels no sentiment or inward impression from this succession of objects: consequently, there is not, in any single, particular instance of cause and effect, *any thing* [sic] *which can suggest the idea of power or necessary connexion* [emphasis added] (Hume 1748 EUH 7.1.50)

In summary, Hume considered causation to be nothing more than observed regularity, supported by observations of the effect not happening in the absence of the cause and irrespective of there being some explanatory mechanism for the events.

This account visibly underpins the desired evidential hierarchies supported by EBHC. We see this as problematic for a discipline which is characterised by complexity and context-sensitivity, as person centred healthcare might be (Miles and Mezzich 2011, 2012).

13.3 A Dispositionalist Response

I will now outline a dispositionalist's response to the emergent problems of a Humean account of causation. I will do this in a framework of four *desideratum* which have been developed from existing commentary on how a causal account for healthcare should look. In order for a causal account for healthcare to be valid, it should:

1. explain the causal role of content from particular research methods;
2. motivate a viable epistemology;
3. account for causal processes in individual level clinical decision making;
4. help understand and assess additional premises and assumptions needed to bridge the inferential gap between population level evidence and clinical decisions.

These can now be taken in turn, with brief attention to the limitations of Humeanism, and an alternative dispositionalist response.

13.3.1 Explain the Causal Role of Content from Particular Research Methods

For the clinician, this simply means that a research theory which we draw causal claims from (i.e. "does exercise work?") should be able to satisfactorily explain *how* and *why* the data from particular methods relates to such claims. Our examples throughout this chapter have been about data from RCTs, and we have seen that the statistical outcomes of RCTs tell us something about the differences between groups, but they do not meaningfully give us an insight into the real causal explanations of that difference. We suggest that a dispositionalist reading of scientific data better explains how it relates to causation.

A traditional Humean account of causation offers some explanation as to how causal claims are developed from research methods relying on statistical data and comparisons of these. Humeans are able to discuss causal claims in terms of either frequencies of occurrence of events, or the degree of differences between two frequencies, or both. Proponents of the Humean account are satisfied that this sufficiently explains the causal role of research content, specifically highlighting that this avoids unnecessary matters of ontology. The dispositionalist response is straightforward: the content that is being referred to here is not of causation, but of something else. The essence of causation has not been reached, and as such any explanation related to causation cannot be given. The truthmaker of causation within Humean accounts is too far removed from where causation itself is most likely to be found.

What dispositionalism offers is a view that sees causation within the core of the content itself: the properties involved in the causal process. Changes are seen within groups, and these changes occur as a result of the interactions of multiple dispositions tending towards and away from effects. Whereas Humeans consider single causes by proxy of frequently occurring observed events, dispositionalists see various causal factors that may or may not manifest in an effect. The causal role of these events for dispositionalism is the notion of how they manifest and how they may tend towards and away from anticipated thresholds. Dispositionalists are unsatisfied with causal explanations that relate to frequentist interpretations of probability, as probability should be thought of in relation to the propensities held by causal factors (see Rocca, ch. 2, this book). (For those who are interested, Donald Gillies (2017) provides a useful primer on the different interpretations of probability on healthcare.) The response here has allowed some appreciation of the way causal content from research methods might be thought of in relation to different ideas about causation.

13.3.2 *Motivate a Viable Epistemology*

Epistemology basically means 'theory of knowledge' and allows us to judge on what grounds our beliefs and opinions are made. So when we say "exercise works for low back pain", we can satisfactorily answer *how we know* this. In healthcare, we want our beliefs and opinions to be as close to the 'truth' as possible. Therefore, any underpinning theory for research and practice must be able provide and encourage an epistemology which any logical, scientifically-minded practitioner could be comfortable with using to justify their beliefs and opinions. We see that there are limitations to a Humean-motivated epistemology which dispositionalism can respond to.

The motivation for a viable and pluralistic epistemology comes from the dispositionalists' commitment to an ontology of the reality of causes, and the understanding that those causes are the most basic and fundamental features of the world. Although others have spoken about multiple methods, (for example, Williamson 2007 and Howick's views on mechanisms in Howick et al. 2013), they struggle to conclude with a convincing epistemology due to their Humean commitments. Framed by epistemological matters, these attempts have been instances of identifying the problems of a traditional monistic account, but trying to resolve these with either pluralistic accounts, or epistemologically driven theories. This is unsatisfactory on two counts: one that there is confusion and conflict between competing theories of causation; and two that in refusing to commit to an ontology of causation, it is not even possible to identify what the methods are searching for.

A theory of dispositions has clearly defined what causation is, and is confident that a methodological pluralistic framework, whereby all methods point to the same thing, offers a satisfactory account of its causal epistemology. A common element in the dispositionalist response across these two desiderata has been the primacy of causation in a dispositions account. Causes are assumed as fundamental and real features of the world with dispositions.

That causes are primitive and real has allowed commentary on both how causal content of methods can be explained, and how a viable epistemology can be motivated. Humeans struggle with explaining causal content because they admit to causation being nothing more than accidental regularity, represented through a frequentist interpretation of probability. Dispositionalists on the other hand take causes as real entities and so can describe and explain precisely their content. As causes can only ever tend towards an effect, dispositionalism does not have to represent causes through frequencies. Rather, a probabilistic theory based on individual propensities offers deeper explanation of the causal content.

Dispositionalism is confident that its visible ontology prepares the ground well for talking about and accounting for a reconceptualised causal epistemology. Its reference to methods such as RCTs being symptomatic rather than constitutive of causation facilitates a methodological pluralist stance whereby information from multiple methods and sources may reveal parts of the causal process. These sources can include indicators of causation such as mechanistic science and patient narratives. Dispositionalists do not need to worry about the reconciliation of multiple theories of causation, because causes are only one thing.

13.3.3 Account for Causal Processes in Individual Level Clinical Decision Making

This is the key issue in science and practice – the problem of induction: how do claims made at a population level relate to individual clinical decision making? Do the outcomes of an RCT relate directly to your patient in front of you? And if so, how do you account for and explain that relationship? A satisfactory theory of causation and research must be able to do just that.

As physiotherapists, we have guidelines which recommend (based on RCT level evidence) that exercise should be considered as an effective intervention for people with lower back pain (LBP) (NICE 2016). This is encouraging on many levels, as there can be many possible co-benefits to exercising, such as improved general fitness and cardio-vascular health. However, a particular patient may have a dispositional make-up which either inhibits them from doing exercise (desire, time, capacity, co-pathologies, etc.), or indeed which tend to make the pain worse during exercise (fear, anxiety, previous experience with exercise, inappropriate loading, etc.). Thus, the person with low back pain is not seen as a discrete and independent variable onto which 'exercise' can be put on to. Rather, this person becomes part of a causal process within which the many dispositional traits of both themselves *and* the undertaking of exercise can be sculpted together to work towards a therapeutically beneficial response to exercise, such as addressing anxieties and fears, coaching on appropriate loading, developing capacity, and so on.

Dispositionalism is a singularist view (Mumford and Anjum 2011: 71–2), and takes particular causal claims and singular instances to be where causation lies (see Anjum, ch. 2, this book). However, the dispositionalist ontology moves quickly beyond a world of discrete events. The *relata* for causal relations are not discrete events or facts – for example, event C (exercise) and event E (improvement LBP) – but the powers and properties of things – for example, the *disposition* of exercise and the *disposition* of a patient with LBP.

At the same time, however, dispositionalism sees general claims as having a role to play in a theory of causation, especially when particular circumstances are not yet known. General claims allow us to be "*armed for future actions*" (Mumford 2009: 14). However, the truth of any general causal claim is substantiated by the properties or dispositions of such claims, and not associations of discrete events and the statistical facts that relate to such. The particular instance, however, that is able to stand separate from a general claim, allows further insight into what causation is.

If causal *relata* are the properties of things, then immediately there is a glimpse of the essence of what causation might be. Causation is now something primitive. That is, it is not something that can be analytically reduced to something else, such as non-causal facts about (repeated) observed associations between events, a difference-maker or a raised statistical probability. Particular instances provide a notion that the modality of a dispositionalist account is neither one of contingency (that is, the probability or possibility of an effect happening), nor of necessity (that the effect will (or will not) happen). What dispositions can say about a particular instance is that there is a modality of tendency that is unique – a *sui generis* (Anjum and Mumford 2018). A cause is something that *tends* towards its effect in a way that cannot be reduced to accidental regularity or necessity.

13.3.4 Help Understand and Assess Additional Premises and Assumptions Needed to Bridge the Inferential Gap Between Population Level Evidence and Clinical Decisions

This point continues to work on the challenge of the "inferential gap". The traditional Humean theory of EBHC can only take things so far and bases its assumptions to cross the gap between general and particular claims on the quality of the research method and frequentist probability. Both are problematic assumptions which are abstract to the world of clinical practice.

Causes are clearly not to be understood as factors that have exactly the same effect in every context in which they appear. Causes that have been identified through RCTs, carried out to perfectly acceptable standards, and clearly suggestive of a certain prediction and clinical intervention, could nevertheless fail to produce their expected effect. When one looks to the ontological matters of causation, one sees that this further consideration, concerning context and composition, can be highly significant. Adding together a combination of drugs, for instance, each of which has been found to have a safe and positive effect in RCTs, could still possibly produce a 'cocktail effect' that is unsafe. Again, this explains why causal inferences are fallible. They are based on an assumption of a finite number of operating factors. An unknown factor could effectively be an additive interferer, for some expected effect. Worse still, it might be a factor that composes nonlinearly with the presence of the other factors to produce an antipathetic effect.

The predictive value of such dispositional reasoning might, however, be questioned by those schooled in (probabilistic) deductive necessity – at least robust methods might have *some* predictive utility within a traditional account, it might be claimed. However, dispositionalism is not relativism and prediction is a feature of dispositionalism. It is not that dispositionalism denies deductivism, although it does judge it to be "*over-ambitious*" (Mumford and Anjum 2011: 140). The difference is subtle but clear: whereas the traditionalist would say 'if A, then necessarily B (to a degree of probability)', the dispositionalist would say 'if A, then B is disposed or *tends* to happen'.

The Humean response is simply to assert that if prioritised methods are conducted correctly – without experimental error – then predictions should be forthcoming that are simple, exact and unfailing. We know this to be false. Any account of causal inferences has to respect the obvious datum that predictions are fallible and defeasible. Dispositionalism offers an explanation of prediction and inference within a fallibilist framework in which dispositions tend to produce their effects but might not always do so.

13.4 Conclusion

We often use interventions with our patients which are based on the *best of evidence*, yet still see no positive response. The Humean might see this as a normal artefact of frequentist reasoning – on average the intervention works, bad luck in this case. The dispositionalist, however, sees it as an example whereby there are insufficient additive causal factors, or too many subtractive factors than desired and so a therapeutic threshold is not reached. The "intervention" is just one possible causal factor, and its success relies on its manifestation with other causal partners. This is what is characteristic about complex and context-sensitive clinical practice. Neither the intervention nor the patient has 'failed', rather, there has not been the anticipated mutual manifestation of variables in this case. Multi-dimensional clinical reasoning frameworks, such as Mitchell et al. (2017), are exemplars of a dispositionalist theory at play. The constantly evolving understanding of how aspects of a person's biological, psychological, social dimension of their life and experience may dispose them towards either an improvement, or a worsening of their health status.

In this chapter, we have learnt something about the relationship between EBHC and causation, and also between theory and methods. Less-than-perfect correlations can indicate something causal occurring, but are by no means irrefutable evidence of some consistent or generalisable causal trend, however strong the correlation. Prioritised research methods can indicate causal processes. However, the causal work is being done within each group and thus it is the groups themselves, not the regularities or counterfactuals, which act as the truthmakers. Robust population studies may be very good at displaying symptoms of causation, but they are not constitutive of causation.

The greatest causal work can be seen in single instance cases. This is where the real nature of causation is witnessed. The interaction between causal agents, subtractive and additive dispositions tending towards and away from an effect, causal powers being passed from one partner to another. For the dispositionalist, the essence of causation becomes apparent. In a dispositionalist ontology, scientific research should focus on the interaction of causal partners and not be dominated singularly by the pursuit for statistical invariance in large groups. For the clinician, the relationship between research findings and individual clinical decisions becomes clearer.

Despite those who reject the utility of anything other than epistemological analyses, an ontological review allows the notion of EBHC to be re-evaluated from bottom-up. One of the foundational intentions of this chapter was to work towards a philosophical underpinning of EBHC that would better support what already happens in clinical practice. This bottom-up approach relates well to clinical practice. Least of all because a dispositionalist account of causation takes the individual patient and the therapeutic interaction, along with all that is known about the process of disease and interventions, as its starting point for an account of causation.

The ontological locus of dispositionalism allows the theory to escape commitments to – and therefore shortcomings of – proxy truthmakers, universal laws, causal necessity, probabilistic inference, and restrictive logical forms. A causal theory based on causal dispositionalism is well positioned to account for the problems represented by the claims of EBHC, and furthermore provides an appealing commentary on a causal epistemological framework. For the shop-floor clinician, a dispositional ontology for causation far better facilitates a person-centred, evidence-informed clinical reasoning approach to best healthcare practice. If this were the only legacy of the CauseHealth project, then it has been a worthy venture.

References and Further Readings

Anjum RL, Mumford S (2018) What tends to be; the philosophy of dispositional modality. Routledge, London

Gillies D (2017) Frequency and propensity: the interpretation of probability in causal models for medicine. In: Solomon M, Simon JR, Kincaid H (eds) The Routledge companion of philosophy of medicine. Routledge, London, pp 71–81

Guyatt G (1991) Evidence-based medicine. ACP J Club 114. https://doi.org/10.7326/ACPJC-1991-114-2-A16

Guyatt G, Cairns JA, Churchill D et al (1992) Evidence-based medicine. A new approach to teaching the practice of medicine. JAMA 268:2420–2425

Guyatt G, Oxman AD, Akl EA, Kunz R et al (2011) GRADE guidelines: 1. introduction-GRADE evidence profiles and summary of findings tables. J Clin Epidemiol 64:383–394

Howick J, Glasziou P, Aronson JK (2013) Problems with using mechanisms to solve the problem of extrapolation. Theor Med Bioeth 34:275–291

Hume D (1740) Abstract of a treatise of human nature. In: Millican P (ed) An enquiry concerning human understanding. Oxford University Press, Oxford, pp 133–145

Hume D (1748) An enquiry concerning human understanding. Oxford University Press, Oxford

Kerry R, Eriksen TE, Lie SAN, Mumford SD, Anjum RL (2012) Causation and evidence-based practice: an ontological review. J Eval Clin Pract 18:1006–1012

La Caze A (2008) Evidence-based medicine can't be.... Soc Epistemol 22:353–370

Lewis D (1973) Counterfactuals. Oxford University Press, Oxford

Mercuri M, Baigrie BS (2018) What confidence should we have in GRADE? J Eval Clin Pract 24:1240–1246

Miles A, Mezzich JE (2011) Person-centred medicine: advancing methods, promoting implementation. Int J Pers Cent Med 1:423–428

Miles A, Mezzich JE (2012) The care of the patient and the soul of the clinic: person-centered medicine as an emergent model of clinical practice. Int J Pers Cent Med 1:207–222

Mitchell T, Beales D, Slater H, O'Sullivan P (eds) (2017) Multi-dimensional clinical reasoning framework. Curtin University Press, Curtin. https://www.musculoskeletalframework.net/

Mumford SD (2009) Causal powers and capacities. In: Beebee H, Hitchcock C, Menzies P (eds) The Oxford handbook of causation. Oxford University Press, Oxford

Mumford S, Anjum RL (2011) Getting causes from powers. Oxford University Press, Oxford

NICE (2016) Low back pain and sciatica in over 16s: Assessment and management. Nice Guideline [ng59]

Sackett DL, Rosenberg WMC, Gray JAM, Haynes RB, Richardson WS (1996) Evidence based medicine: what it is and what it isn't – it's about integrating individual clinical expertise and the best external evidence. Brit Med J 312:71–72

Williamson J (2007) Causal pluralism versus epistemic causality. Philos Mag 77:69–96

Open Access This chapter is licensed under the terms of the Creative Commons Attribution 4.0 International License (http://creativecommons.org/licenses/by/4.0/), which permits use, sharing, adaptation, distribution and reproduction in any medium or format, as long as you give appropriate credit to the original author(s) and the source, provide a link to the Creative Commons license and indicate if changes were made.

The images or other third party material in this chapter are included in the chapter's Creative Commons license, unless indicated otherwise in a credit line to the material. If material is not included in the chapter's Creative Commons license and your intended use is not permitted by statutory regulation or exceeds the permitted use, you will need to obtain permission directly from the copyright holder.

Chapter 14
The Practice of Whole Person-Centred Healthcare

Brian Broom

In the spirit of being whole person centred I will start with a story (some details are changed or not included to ensure confidentiality):

14.1 A Woman with Skin Disease

Recently I was consulted by a woman with a 35 year history of a disfiguring skin condition complicated by a crippling arthritis. She had sought help from many practitioners, orthodox and unorthodox. Despite much medical treatment it remained out of control. She admitted quite spontaneously that she had moved quickly from one promising healing modality to another. She asked for help from me because she had read in a magazine that I looked at the 'whole person'. The unaddressed story that unfolded in our meeting began in her family of origin. She was conspicuously clever. In contrast, her sisters were more oriented to traditional domestic roles. Her father could appreciate the sisters' practical skills but could not validate her inclinations and academic achievements. As a young adult she spread her wings and travelled widely. Again in contrast, her sisters were all mothers by the age of 21. In her late twenties she married and conceived. The comment in the wider family was "Oh my God, she is going to have a baby!" The sense of inadequacy and self-doubt was steadily accumulating. The birth of her baby was a disaster: a very prolonged delivery in a remote setting, a genuine risk of both baby and mother dying, the baby with the cord tight around its neck and snatched away after delivery for resuscitation, the baby not held for two days and ensuing poor bonding, and a crushing sense of powerlessness and loneliness. She struggled with the child for years and while he was still an infant the father left the marriage for one of her friends. She blamed herself, her inadequacy, for this. The skin condition began and has continued ever since. She

B. Broom (✉)
Auckland University of Technology, Auckland, New Zealand

noted that it got much worse after another of her children suddenly died of unknown cause. Another child has developed a different serious inflammatory disorder. One way and another she blames all this on her inadequacy. There were some hints of very early sexual abuse. Notably, she described herself as struggling with something "ugly and dark" within her.

The question of course is why would I bother to take out such a history in a person with a serious medical condition? My reactive response is simply this: why not? A more considered answer is that the story seems really powerful and that from a unitive, nondualistic, whole person-centred perspective, in which mind and body are not separated in the ways we have traditionally accepted in medicine and culture generally, it is very likely these story factors are playing a significant role in her suffering and disease.

14.2 A Professional Evolution

In 1982, in mid-life, I abandoned a flourishing academic clinical immunology career, at the Christchurch School of Medicine, to train in psychiatry. I was responding to an increasing sensitivity to fragmentation. I had been reared within a Christian spirituality largely unaffected by science and modern thought and scholarship. I had trained to high levels in internal medicine which largely ignored subjective experience. And I was yet to discover a psychiatry and psychotherapy largely ignoring the body. I felt that this systemic fragmentation was a fundamental cultural problem, but had little ability to articulate it, let alone convert it into clinical practice.

Thus began a journey specifically undertaken to explore relations between medical practice and patients as whole persons. I had the impulse but no real concept of what it meant and certainly no idea where I was heading. I was not popular. My highly esteemed mentor, senior colleague, and head of the department of medicine, Professor Don Beaven, gave a simple and direct response to this change in direction: "You are an idealist!" and of course he was right. Other colleagues and peers seemed to feel I was betraying an unwritten professional code. The forces maintaining normative cultural structures in medicine are very powerful.

Already having expertise in the body (or at least, *diseases of the body*), my first move was to embrace the 'mind'. Entering psychiatry allowed me to begin this process, to maintain my medical functioning, and to earn a sufficient living to sustain my growing family. The four years in psychiatry taught me many interesting and useful things, but I found that as a medical discipline it was pretty much just as deeply embedded in physico-materialist and dualistic assumptions as my previous internal medicine framework. This was not what I wanted. I was seeking to understand *persons* and the potential for integration of all dimensions of personhood in our understanding and treatment of illness and disease.

Psychiatry does have an interest in those physical illness presentations known as *psychosomatic* conditions, but had long vacated interest or professional responsibility for mind factors in all the other physical conditions, the 'real' (sic) physical diseases. Indeed, mostly, it was felt that the apparently *non-psychosomatic* conditions had nothing to do with mind or subjectivity.

Psychotherapy, on the other hand, held more promise, and so I veered away from psychiatry. New 'worlds' of thinking opened up cumulatively over many years, about the mind or the subjectivity of persons as patients. The most influential emphases were psychodynamic theory, infant development, stages of life concepts, family and systems theory, trauma concepts, object relations theory, interpersonal psychotherapies, self psychology, learning theory, narrative theory, and consciousness studies. These worlds had been entirely invisible to me previously as a clinical immunologist. They remain invisible to the vast majority of medical clinicians working in practice in the Western world. The point I make here is that there is a vast panorama of the subjectivity of persons excluded from the ordinary arenas of medical care.

Psychiatry was not a suitable base for further exploration. I initiated a multidisciplinary Centre (Arahura Centre, Christchurch, New Zealand) committed to the integration of high standard medical practice, psychotherapy, and spiritual values. All of the staff and trainees came from diverse Christian backgrounds and felt similar if not identical aspirations for integration. With two colleagues I mentored this unusual journey of integration. Over the next ten years what developed was a multi-dimensional, multi-factorial, multi-causal, and multi-methodological approach towards disease (vide infra).

Personally, I began a psychotherapy practice and re-ignited my role as a clinical immunologist. There was nothing particularly intentional or inspired about that decision—it just seemed a sensible way to continue my life as a clinician. But, to my surprise, startling and jolting things emerged. Before entering psychiatry I had enacted my clinical life largely by perceiving physical diagnoses and diseases. Now, as both a physician and as a psychotherapist, working with individual patients presenting with a wide range of physical conditions, I was still making diagnoses and treating diseases **but also** hearing 'stories' in the same clinical time/space. And I started to see connections between diseases and stories in many cases. Thus began my work with 'Medicine and Story' (Broom 2000).

14.3 Somatic Metaphors

What was more disturbing was that these stories suggested that some of the physical conditions I was treating were actually *symbolic*. I have written extensively about these (2, 3). I called these instances *somatic metaphors* (Broom 2002). There were sexually abused patients with oral and genital conditions. A patient with a facial rash keeping a 'brave face' on a partner's depressive condition *and unable to talk about it*. A patient with years of crippling mouth ulceration resolved when she finally talked to her daughter about leaving the Catholic Church. My books relate a myriad of such examples, in both relatively minor and also very severe and serious physical disorders, most of which could not be confined and dismissed as *psychosomatic,* in the old sense of the term. More than that, the stories couldn't be simply dismissed as retrospective, narrative constructions or interpretations. The stories appeared to be triggers of the disease. Many were very chronic and had failed to respond to biomedical therapies, but got better when these meanings were ascertained and worked with.

This was very challenging. Initially a major issue was scientific plausibility. How can very specific meanings get expressed in the body, emerging as a symbolic disorder? The dualistic model we are trained in treats the mind (with its meaning-making) separately from the body, in which meaning has no place or role in the aetiology and pathogenesis of disease. In the end, I had to toss this model away and start to think of persons as wholes. The dualistic model itself becomes the implausible construction of reality. Thankfully, in recent years this is becoming more mainstream, at least in wider culture, though it hasn't dawned significantly on medical practice.

14.4 Whole Persons in the Clinic

While I found all of this both exciting and difficult conceptually, the next important issue was how to talk to patients about mind and body connections. What I found was that if I was skilful and did not psychiatrise these connections, patients were by and large cautiously open to a multi-causal view of disease that included 'story', or subjectivity in general.

Instinctively most people know we are 'wholes', it is just common sense. Nevertheless, one reason why we may express our life struggles in physical illness is that we may be unable, for one reason or another, to find a better way to represent, express, and work through certain difficult or painful emotions and issues. But I gradually learned to educate, to warmly 'hold' people through their uncertainties, and to enquire in such a way and at a pace which enabled trust and safety to flourish.

Thus, skills were needed, beyond the algorithmic protocols of normative medical interrogation. A majority of patients love being treated as persons with stories, as opposed to just being diagnostic challenges and objects. We clinicians need to make room for these stories, and, contrary to what most clinicians assume, this does not need to be principally about time cost. Curiously, it is much more about generosity, safety, empathy, and simple information and education about mind and body connections (see www.wholeperson.healthcare for an expanded review of appropriate listening skills). But until clinicians accept that human subjectivity or stories play a role in disease development none of this is going to be prioritised.

14.5 Reactions from Colleagues

Another issue was how to communicate with my colleagues. Patients were referred to me as an immunologist for *immunological assessment*. I wrote report letters back to my colleagues which certainly offered that, but also, where relevant (and almost universally), the review was wrapped in or permeated by a story.

I was anxious about this initially. How would I be seen? Would referrals dry up? My practice flourished. Many family practitioners welcomed these more fulsome reports. The odd one was dismissive (usually reported by a patient). Many carried on as before, impressed by but seemingly unaffected by my educative reports. Increasingly doctors sent me their problematic patients—the people who did not respond to standard biomedical care. I became known as a doctor who could hold complexity. Some doctors embraced the approach and trained with me.

But I learned early that despite even overwhelming evidence of mind and body connectedness most clinicians want to stay with what they know and do, with their dualistic biomedical model, even if this is not in the best interests of their patients. As an example of this, there are a host of scientific psycho-neuroimmunological articles suggesting an impact on the immune system of psychological factors, stress or abuse. And yet again and again I have attended huge immunology conferences where there is clearly no interest in or evidence of any impact of this work.

This problem of the neglect of the role of the mind or subjectivity factors is both multifactorial and formidable. The issues include: vested interest based on training, time and income flow; the dominance of the biomedical model; default behaviours and skills that militate against listening; lack of psychological understanding on the part of clinicians; relational inadequacy; inability to cope with patients' emotions; financial, institutional, cultural and systemic structures that avoid human experience; the valuing of quantitative over qualitative evidence; and the difficulties of dealing with data that is not easily 'measured'.

14.6 Dualist Psychotherapy

But we pressed on. The numbers coming to me increased and I found myself needing psychotherapists to help me. I would assess the patients and refer them on to excellent community psychotherapists. A new phenomenon appeared. Psychotherapists are good with 'stories' but they often, by default, exclude the 'body' from their clinical working space. This would happen even if I sent a patient specifically for 'mindbody' psychotherapy for a physical condition like urticaria, eczema, asthma, irritable bowel syndrome, migraine and much more. I also discovered that the patients would not do as well as I had expected. On review, I found these patients told me that they might introduce their physical symptoms in the session with the therapist, who then would typically ask if the patient had talked to the family doctor about it again. That is, essentially, the therapist was saying to the person, 'take your body to the doctor, it is not my job.' Psychotherapists are certainly conceptually more open to the mindbody connections, but in practice are as dualistic as doctors. So I established supervision groups for psychotherapists, and from that time only ever referred my patients to the eight therapists I had in mindbody supervision. The therapists themselves needed holding in a whole person-centred framework. The outcomes were much better.

The point is of course that all of us in our specialised clinical work are dualists to some degree. Professionalism and scopes of practice serve to keep us in our conceptual silos. Again and again I find myself reiterating the two bulwarks of whole person healthcare work. We need to have a non-dualistic view of persons and disease, and we need to have the skills to comfortably allow mind and body to be *together* in our clinical workspace with patients, and to listen and respond. We now have many clinicians who can do this. They do not need to be dually trained (Broom 2013), as I am in both medicine and psychotherapy. It helps greatly if different disciplines are in close contact, supporting (for example) the doctor towards stories and psychotherapists towards bodies.

14.7 Publications

Time went by, and eventually in 1997 I published my first book, *Somatic Illness and the Patient's Other Story* (Broom 1997), essentially to tell the story of physical illness as related to the predisposing, precipitating and perpetuating 'story' factors of causality, especially focusing on those diseases generally regarded as physical, but also those considered as psychosomatic. The principles of non-dual, unitive, whole person practice apply of course to all illnesses. The book provides many examples of the observed clinical phenomenology, and lays out in detail the skills necessary to provide treatments that actively integrate normative biomedical principles with story factors.

In 2007 I published a second book, *Meaning-Full disease. How Personal Experience and Meanings Cause and Maintain Physical Illness* (Broom 2007), systematically addressing and arguing a theoretical basis for seeing persons as wholes, understanding diseases as being expressions of wholes (that is, multifactorial, multidimensional and multicausal), and the benefits of treating them as wholes. In that book I drew on many resources and concepts that can contribute to such discussions. Once we escape the grip of strict scientism we can find help from many disciplines, such as philosophy, modern physics, complexity theory, cultural and trans-cultural studies, the arguments for and results of qualitative research, and much more.

But clearly the biopsychosocial model (BPSM), psychosomatic studies and insights, and psychoneuroimmunology (PNI) deserve brief further mention here. I have gained a lot from these perspectives, but I diverge from them in one crucial respect. All are permeated with a fundamental dualism or physicalist reductionism or both.

My view is that George Engel, the originator of the biopsychosocial model, was not particularly dualistic, but the vast majority since who have espoused the model are indeed so. I really tire of hearing clinicians claiming to *believe* in the biopsychosocial model, but essentially happily ignoring the role of psychosocial elements in the supposedly 'real' physical diseases.

For its part, the psychosomatic tradition keeps mind-oriented clinicians focussed on a small group of disorders that cannot be easily *explained* by biomedicine—as if subjectivity is really only relevant in this small grouping of disorders, which cannot be explained by the biomedical model. This is rampant dualism.

Psychoneuroimmunology presents a different story again. It is as if we can only accept a role for story if we can find a linear mechanism acceptable to the biomedical model. Therefore it is fundamentally body-focussed and reductive.

One might also think that narrative medicine and narrative therapies have a lot in common with my story approach. Essentially they arose out of the 20th Century post-modern ethos and cross-cultural studies. The narrative focus is important. Essentially it is about what sense people make of their illness. This is important but it is a sub-section of story-gathering. Story in the narrative sense is essentially *post-hoc*: that is, the physical disease arises because of purely physical factors and *then* is interpreted *after the event* in one way or another, according to one's belief systems and cultural influences. Our work goes beyond that to understanding the role of story in 'causing' disease.

14.8 Human Infant Development

Let us consider a human infant in its earliest phase of development. What kind of concept do we have of that infant's evolving dimensions of physicality (bodily development, growth, movement, coordination etc) and subjectivity (capacity for experience, perceiving, relating, thinking etc). Surely it is without question that these capacities emerge and evolve from the beginning, *together*. The subjectivity dimension is not some adjunct, latter-day appendage, a kind of discretionary item. The dimensions we call body and mind co-emerge, and are inextricably integrated. I and two colleagues have addressed this in a paper titled *Symbolic illness and 'mindbody' co-emergence. A challenge for psychoneuroimmunology* (Broom et al. 2012).

If mind and body are not inherently separate, why would we assume that, in disease, body is important but mind is not? The reasons given for this depend a little on one's background, but include the sharp 'turn' to dualism attributed to the French Catholic mathematician/philosopher, Descartes, or the inherent dualism of language-making, or the primacy given to concrete measurable dimensions of reality, or the fears we have of revealing the harsh and vulnerable aspects of our personal experience, or the determined 'disenchantment' of the world during the last few centuries in the name of demystification, positivism, and mastery. Most likely all these have played a role.

But our whole person-centred phenomenology of disease suggests strongly that we must engage in a unitive view of persons as wholes, and treat diseases within that conceptual framework.

14.9 *Mindbody* Healthcare

Back to what that meant in practice for me. I worked at the Arahura Centre from 1987 until 2007. In the later years I initiated (2005) a multi-disciplinary Masters and Diploma Post-Graduate Program in MindBody Healthcare at the Auckland University of Technology, specifically aimed at experienced clinicians of all kinds (for accounts by the clinicians themselves see Broom 2013). I had long felt that what we must do was encourage clinicians to expand their practices to include a story element. I had done many well-received workshops and seminars, nationally and internationally, but became aware that for most people it was too hard to make the changes needed. Once clinicians accepted and embraced the concept of persons as wholes, they typically became much more aware of what they were not doing. Some wanted to apply it to everyone and soon became overwhelmed. Many did not have good listening capacities, did not know how to invite patients to reveal difficult things, were uncomfortable with emotions, feared running overtime, and simply didn't have a language for these kinds of conversations. Some had habits that they either couldn't or did not want to break. Some would try, and then run into trouble based on these issues. So we devised a program which got established at the University.

The multi-disciplinary side worked out very well. It meant everyone had to see the generic principles of conceptual understanding of wholes, multifactoriality and multidimensionality, and also see that the skills of listening to story were the same whatever discipline we were in. What was different was how this worked out in different disciplinary settings.

Commonly we saw a lot of personal confusion and incoherence developing for more than 50 percent of the students in the first few months as they started to transform their behaviours and practices. There are typical issues. How do I start a story conversation without alienating the patient? What do I do now that the story has come out? Learning not to jump to *fixing* things was a major hurdle for the body clinicians (doctors, physiotherapists). Understanding that most stories are about what and what has not happened in relationships was hard for some to appreciate. Some would jump at a meaning like a clinician would jump at the result of a CT scan, and then find the patient pulling away. Being person-centred rather than 'my expertise'-centred was a harsh lesson for some. Discovering that collaborative recognition, between clinician and patient, of just the fact of a connection between the presenting illness and the story might be in itself enough to settle an illness was an amazing experience for others. Wanting it to be that way every time was a salutary lesson for others, whose desire for mastery transcended their capacity for tolerating complexity.

We have trained some 70–80 clinicians in this way.

In the earlier years I commuted to Auckland to teach this program but in 2008 I shifted there to live. I continued the AUT program, and was invited to take up a position as an Immunologist at Auckland City Hospital. It was then 26 years since I had initiated and led the Immunology department at Christchurch Hospital. I wondered how I would fit, given all the developments in my thinking and practice over the years. It was nice to be back in an old familiar environment. I was determined not to badger people into my way of thinking, and more or less succeeded in that. Throughout my time there I continued to practice from a whole person perspective. This was respected and valued. It has always been important to me to be an excellent physician in a normative sense, and being amongst biomedically-informed colleagues helped me keep up that aspect of my functioning. For years now, we have had two to four Masters level students from the Department of Psychotherapy at AUT University on placement in the Immunology Department working with Immunology patients and supervised by myself and a very experienced MindBody Psychotherapist from AUT.

14.10 I Was Conflicted

But I often felt heavy after my days at the hospital, constantly aware of patients who were struggling because of inadequate purely biomedical approaches, or suffering unnecessary side effects of treatments, and who could have been helped by a whole person approach. I was conflicted. I am very fond of, and greatly enjoyed the company of my colleagues, but at times I became impatient and frustrated that they could/would not open themselves a bit wider for their patients. There was a genuine respect for what I was doing, they wanted me there, but generally they hesitated to do it themselves. Some things did change in the department. Most of the clinicians professed to be affected by the whole person approach, and there was a reduction in useless and pointless testing for many patients. But it didn't go far enough for me. Perhaps it was the idealist in me, wanting a more major re-orientation. One of my colleagues has an enthusiastic and competent grasp of the approach. And we have 14 specialist doctors from throughout the hospital who meet monthly to talk about and learn the approach.

We are truly faced with a conundrum. There is certainly a hunger for something different, more whole person-oriented, but there is a powerful inertia in the healthcare system.

There is one thing more for this chapter. What is the cause of improvement in illness and disease when the whole person-centred approach is used? As stated at the outset, a major stimulus for me as a clinician was the phenomenology of symbolic disease. Thus a focus on meaning and stories of meaning-full disease became my initial doorway into the 'whole person'. But I have resisted a reduction of all disease to meaning. Opening things up by going through the doorway of meaning mobilises many other elements, including the importance of relationship in healing.

The skills involved in listening to stories are mainly relational skills. Genuine interest, willingness to tolerate uncertainty, waiting to see what emerges rather than relying on prior knowing, warmth and empathy, and much more. Certainly not all patients had symbolic disorders. Most stories we dealt with were crucially about relational disturbance.

This leads to another issue. Clinicians, by virtue of their training, are very susceptible to learning a *methodology* that is based in searching for a story, or *the* story, like searching for a diagnosis or a pathogen. But once such a search is initiated they are confronted with what it means to be in a healing relationship with the patient in a much more intimate way.

Out of all of this, I came to suspect that much of what we were doing could be conceptualised roughly as follows. Meaning is important, and it is a physician's focus in opening up the field, in which it is accepted by the clinician and patient that mind and body are connected and need to be addressed in the proper treatment of a physical condition. Physician interest in the patient's meanings or story triggers in the patient a sense of being understood. Sufficient trust and safety can lead to the patient revealing painful and important material, and with support be nudged into resolving some long-standing issues, usually of trauma or disturbance in relationship. It might be enough just to name these, or a longer process of mindbody psychotherapy might be needed. At every point physical and non-physical factors are held to be potentially contributory.

All this might be the physician's interpretation of what is happening. But what do the patients say?

14.11 Being Looked at or Being Seen?

Galia Barhava-Monteith, a mature woman who had recovered from the serious disorder Churg-Strauss Vasculitis whilst undergoing both normative chemotherapy and a whole person-centred approach, recently completed a PhD (AUT University 2018) exploring this question, titled: *"The difference between being looked at and being seen": An in-depth consideration of experiencing the Whole Person Therapeutic Approach for chronic illness"*. There are many important things to ponder on in this thesis. It is noteworthy that, in contrast to the perspective of whole person-centred clinicians as to what is important in disease treatment, few of the patients undergoing such treatment refer to the importance of specific meanings or symbolism in their stories of recovery.

It will suffice to use Dr Barhava-Monteith's own words:

> I have come to see that the taken-for-granted practices of WPTA (*Whole Person Treatment Approach*) clinicians, or the *how* of their practice, is the thing itself. I now comprehend how the therapeutically beneficial aspects of WPTA, and any other encounter between two people where one's role is to enable the other to get better, are dependent on the capacity of the

healer to see the person they are trying to heal. The technical knowledge of course must be there, but it is only the starting point. I came to conceive that this experience of being seen is the feeling that someone else is truly seeing the whole of you. Risking hyperbole, I argue that this is a profoundly existential experience, as mostly we spend our lives with people who only look at aspects of us. Who look at our professional background, or at our symptoms, or look at our childhood trauma. From apprehending that the experience of being seen is somehow important, I now comprehend that it is an existentially humanising and healing experience that unfortunately too few people experience. My doctoral research has confirmed to me that being introduced to non-dualistic concepts as they pertain to health and wellness in the context of this relationship can be transformative, in that patients come to reconceptualise their identity with respect to their illness so that they, like me, can experience freedom and hope. However, I now grasp that to introduce radical new ways of thinking about one's self is something that should be treated with utmost care and reverence. Introducing such notions may destabilise the very core of one's personhood. Now I see that anyone who is involved in activities that are concerned with changing the way another person thinks and behaves, has to earn the right to do so. And the ways by which we earn that right is through being careful and mindful of our words, our actions and our capacity to recognise the personhood of another, and of being experienced as doing so. In a sentence, my conclusion at the end of my doctoral journey, is that it is through the *how*, not the *what* where profound experiences and changes occur… In concluding this doctoral thesis, I do think that making explicit the ontological dualistic assumptions underpinning much of the teaching of modern medicine is important. However, I now think that we need to consider shifting the almost exclusive focus on content or the *what* of clinical sciences, to include, as equal, the *how*, the ways of acting and being. My reflection in concluding this thesis is that clinicians who work to embody these humanising practices, are likely through doing and experiencing, to shift their own preconceptions about the dualistic nature of health and illness. (Barhava-Monteith: 230)

I finish as I began. In considering causality in illness and disease, our work draws attention to the fact that we, as patients, are persons, but more than that, persons-in-relationship. Any therapy that reduces us to a more limited view of persons, such as objects to be technologically manipulated, is going to have serious limitations, if not in many cases profoundly inadequate.

References and Further Readings[1]

Barhava-Monteith G (2018) The difference between being looked at and being seen: an in-depth consideration of experiencing the Whole Person Therapeutic Approach for chronic illness. Dissertation, Auckland University of Technology

Broom BC (1997) Somatic illness and the patient's other story. A practical integrative mind/body approach to disease for doctors and psychotherapists. Free Association Books, London

Broom BC (2000) Medicine and story: a novel clinical panorama arising from a unitary mind/body approach to physical illness. Adv Mind Body Med 16:161–207

Broom BC (2002) Somatic metaphor: a clinical phenomenon pointing to a new model of disease, personhood, and physical reality. Adv Mind Body Med 18:16–29

---

[1] *For clinicians and others wanting ready access to a range of resources in Whole Person-Centred Healthcare there is a wealth of material at* https://wholeperson.healthcare

Broom BC (2007) Meaning-full disease: How personal experience and meanings initiate and maintain physical illness. Karnac Books, London

Broom BC (2013) Transforming clinical practice using a mindbody approach. A radical integration. Karnac Books, London

Broom BC, Booth RJ, Schubert C (2012) Symbolic illness and 'mindbody' co-emergence. A challenge for psychoneuroimmunology. Explore 8:16–25

Open Access This chapter is licensed under the terms of the Creative Commons Attribution 4.0 International License (http://creativecommons.org/licenses/by/4.0/), which permits use, sharing, adaptation, distribution and reproduction in any medium or format, as long as you give appropriate credit to the original author(s) and the source, provide a link to the Creative Commons license and indicate if changes were made.

The images or other third party material in this chapter are included in the chapter's Creative Commons license, unless indicated otherwise in a credit line to the material. If material is not included in the chapter's Creative Commons license and your intended use is not permitted by statutory regulation or exceeds the permitted use, you will need to obtain permission directly from the copyright holder.

Chapter 15
A Broken Child – A Diseased Woman

Anna Luise Kirkengen

> *We are such stuff*
> *as life is made of –*
> *and our lived body*
> *tells its tale.*

15.1 Cecily Cramer

Now in her mid-forties, Cecily Cramer[1] is a well-educated, married mother of three children under 17-years of age. My initial contact with her came at the suggestion of her therapists, psychologist Aina, and psychomotor physiotherapist Sanne, both of whom are highly experienced, trauma-oriented professionals.

In our correspondence prior to our first meeting, Cecily authorised me to access the medical records from both of her recent university hospital psychiatric ward admissions, as well as the notes from her out-patient clinical care following the second of these. Cecily also gave me her consent to read the notes from her sessions with Aina and Sanne. Lastly, she allowed me to see the detailed results of an x-ray examination that the university hospital's radiology department had carried out, requisitioned by the hospital's oncology department.

As Cecily and I discussed this documentation in depth during our conversations, I became convinced that her complex sickness history would remain incomprehensible, both to those treating her and to Cecily herself, were her painful life history not taken into account. In fact, her current symptoms could *only* be decoded once their trajectory had been traced back to when they first emerged: at that point, they

---

[1] Cecily Cramer (pseudonym) has contributed to the present form of her history and consented to its publication.

A. L. Kirkengen (✉)
Department of Public Health and Nursing, Norwegian University of Science and Technology (NTNU), Trondheim, Norway
e-mail: anlui-k@online.no

were the sole survival strategies available to her. They did aid her, initially, in her struggle to protect herself. Over time, however, her continuing to rely on those same strategies became a dangerous habit, one that depleted her vitality and literally put her life in jeopardy.

When Cecily recognised the pattern and its impact, she realised that she both wanted to, and could, seek help to break free of it. Her 4-year cooperation with Aina and Sanne enabled her to do just that, and, in the process, regain her self-respect, increase her confidence, improve her health, and help her find the strength to go on, both with her personal life and with her profession.

15.2 Crisis Onset

At 20 years of age, Cecily was diagnosed with a rare, aggressive cancer requiring intensive treatment, including surgery, medication and follow-up. After some years, she was pronounced free of cancer. She married, worked full-time as a specialist nurse, took part in competitive sports, and gave birth, each time with complications, to three healthy children who all developed well. Then, her cancer recurred. Again she went through a difficult course of treatment, and again was declared cancer-free. Toward the end of the long sick leave that followed her treatment, the new leader of her department at work, whom she had never met, began urging her to return to her demanding position, full-time. Cecily feared she would lose her job if she refused. Although she did not feel ready, particularly with three small children at home, Cecily "obeyed" and went back to work. She even agreed to work shifts during Christmas.

By January, she was feeling suicidal. She felt overwhelmed by hopelessness, suffered from frequent nightmares, anxiety, insomnia and an inability to concentrate. Desperate and scared, she was admitted to the psychiatric ward.

15.3 Two In-Patient Psychiatric Hospital Ward Admissions

Records from her first admission read as follows:
Woman with known depression, admitted by her GP due to increasing depressive symptoms. Allergic to penicillin and a possible adverse reaction to another drug. Predispositions: alcoholic father, cousin with personality disorder.

No one, however, asked Cecily: How were you affected by your father's alcoholism? How would you describe your parents? How did they treat you? What do you know about your cousin's condition?

Records:
Somatic: surgery and treatment for kidney cancer, 21-years old.

15 A Broken Child – A Diseased Woman

No one, however, asked her: How did that serious disease, and the long-term treatment for it, affect you and your life, especially considering how young you were?

Records:
Patient denies having a substance abuse problem.

No one, however, asked her: Have you ever been abused or maltreated?

Records:
Psychological: Prior depression 2 years ago treated pharmaceutically; considerable side effects from the first meds. Second meds, ineffective.

No one, however, asked her: What would you say is the main reason you feel hopeless and down?

Records:
Patient felt she needed to return to full-time work shortly after treatment for recurring cancer because she was afraid of being fired.

No one, however, asked her: Did you feel pressured, or forced, as if someone had control over you? If so, might that have reminded you of something that happened to you earlier?

Records:
In addition, patient has started to see a psychologist, which has opened up old problems (among these are childhood incest) that the patient, as she puts it, "has trouble coming to grips with".

No one, however, tells her: Nobody ever experiences abuse without being deeply marked by it. No one asks her: What do know about how these early experiences of being abused have affected you and your life – and what do you remember about whoever abused you?

Records:
The patient reports that she has lost her appetite, doesn't sleep properly, doesn't want to get up in the morning, that her memory fails her, that she can't concentrate and suffers at night from anxiety and nightmares.

No one, however, suggests: Your ailments seem to be part of a pattern, perhaps connected to having been under too much stress for too long a time. No one inquires: Do you see any relationship between these problems and experiences from your childhood?

Records:
The patient says that she wants to commit suicide, sees it as her last way out, but hesitates out of concern for her husband and her three children.

No one, however, asks: Why can't you, or why don't you want to, go on with your life?

Records:
The patient confirms that she has had such periods of depression throughout her entire life.

No one, however, asks: Might your feelings of powerlessness and exhaustion have something to do with what was done to you when you were a child? Was there anyone who protected you?

Records:
The patient appears to be suffering, but there is no indication of hallucinations or psychosis. Conclusion: known depression, increasingly aggravated by workplace problems; suicidal ideation. Treatment plan: mood stabilising followed by further treatment.

Addition to the records made by Cecily's ward psychiatrist the following day:
She describes a personality structure characterised by not wanting to speak out or object. She has been eager to do her best, which has probably contributed to her load becoming heavier than necessary. She is sad, prone to weeping, feels hopeless and her thinking is chaotic. Her depressive symptoms range from moderate to high. Her personality does seem to predispose her to depression, although current external stressors have also contributed.

Here, she is assigned a diagnosis while, simultaneously, being defined as its origin – her predisposing "personality". No one seems to have been listening to her when she spoke of having an alcoholic father, or of a childhood marked by incest, as well as other problems – all of which she says she is struggling even to grasp. Psychiatry seems totally deaf to this alarming and highly relevant information.

Addition to the records made by the psychiatrist in charge of Cecily's case, one week later:
The short-term aim for treatment is for her to become calm and stabilised. The long-term aim for treatment is to discharge her in an improved condition.

Nothing in this plan would indicate any intention of seeking to understand the patient, or of offering to help her come to grips with her basic problems. The record does mention, though, that Cecily's medication is to be increased to eight psychoactive drugs.

Addition to the records made by the psychiatrist after 10 days of in-patient treatment:
*The patient still shows depressive symptoms. She finds it difficult to participate in planned activities. Her thoughts are in turmoil, constantly circling around the past; in today's conversation **we** try again to talk about it being natural to think of the past when depressed, so that it soon fills all the space. Therefore, **we** ought to try to concentrate on the actual here-and-now and work with the depressive elements, in addition to acquiring tools to think more positively.* [Emphasis added.]

15 A Broken Child – A Diseased Woman

The psychiatrist uses the plural: "…**we** try…**we** ought". This **we**, however, does not include the patient. By using this **we**, the physician nullifies the patient's stated need to come to grips with what disturbs her thoughts: her past. He shows no interest in this past, nor any intention to inquire into it. He has defined the problem as "here-and-now" with no relationship to "there-and-then".

Four weeks into Cecily's hospitalisation, she meets with psychologist Aina for the first time. Only now is she invited to recount memories of her traumatic childhood and of her inability to comprehend her own emotions. She tells Aina about her younger, multi-handicapped brother for whom she felt responsible because he was so vulnerable, describing herself as "defenceless". She shares that she does feel safe on the ward but misses her children. She admits that, even after 20 years of marriage, she still keeps secrets from her husband, and she reveals to Aina that she starved herself as a teenager because she wanted to die.

Cecily meets with Aina six times before being discharged. Aina summarises in her notes that Cecily has been traumatised, unseen, neglected, and under exceptionally high stress for a prolonged period. Nonetheless, she avoids making any demands on the people around her.

Addition to the records made by Cecily's psychiatrist the day before she is discharged:
The patient is very concerned about whether to talk with somebody about how she was earlier in life regarding all the stress she has been under. We discuss that it is certainly important that everything be dealt with, whenever she is ready. As things are now, however, we must take one thing at a time. Her depression needs to be the focus, and for her to feel calmer and not burden herself with too many stress factors.

The psychiatrist does admit that it would be important, eventually, to discuss the past, although he does not explain why. For him, the disorder itself, *depression*, represents a greater threat to the patient than does the *source* of the disorder. Thus, he separates the disorder from its origins, which he terms "stress factors", without having made any attempt to learn what those might be.

Upon discharge, Cecily is prescribed **seven psychoactive medications**. The physician recommends a prolonged sick leave before a gradual return to work. She is promised out-patient clinic follow-ups to begin immediately, but none is ever offered to her. Despite it being clear that Cecily is on extended sick leave, her employer continues to urge her, insistently, to come back to work.

Six weeks after being discharged, Cecily is readmitted to the same psychiatric ward, in even worse condition than at her first admission. Again, she states that she cannot bear entreaties, demands and threats from someone who has power over her. Again and again, she speaks of how terribly her boss has treated her, of how she has been abandoned by her employer and by the healthcare system that "forgot about" her after her recent discharge.

This time, Cecily meets psychologist Aina after only five days, but they manage to have just two meetings during Cecily's 18 days of hospitalisation. She is told to continue taking four medications for depression, one for restlessness, two different analgesics to be taken as needed, one drug for nausea and one sleeping pill.

15.4 Follow-Up Care

Without explanation, Cecily is referred to the out-patient clinic for those with personality disorders for her follow-up care rather than, as had been planned earlier, the clinic for people suffering from mood disorders and depression. She is immediately offered an appointment with Aina, however, and from then on, the two meet and talk on a regular basis.

When asked how Cecily felt being on the ward, she responds:

*The ward nurses and the team in charge of my treatment were told not to talk to me about my past but to keep the focus on my depression symptoms. I felt ignored. Wrong. Put in a box for depressed women. The conversations were only about what **they** thought was important and about giving me psychological education. Everything was about what I myself could do to overcome my need for isolation and loneliness. But this was exactly what I'd been trying to do on my own, before being admitted. The only advantage to being on the ward was that I didn't commit suicide. And I met Aina, and she had the courage and strength to object to "pre-packaged treatment" and take me under her wing.*

With Aina, Cecily begins to share information about her life that she had not shared before, including, after a few meetings, about her fear of her father. The details that psychologist Aina is enabled to glimpse about what Cecily had been subjected to as a child, convince her to suggest that they involve psychomotor physiotherapist Sanne in the treatment process. Cecily agrees. Sanne examines Cecily's body, her posture, musculature and her patterns of pain and tension. A dual therapeutic process then begins, with sessions twice a week. As often as possible, Cecily's meetings with the two therapists are scheduled to take place on the same day, sometimes with all three working together.

The issues and specific findings that are uncovered are nowhere to be found in Cecily's extensive records, neither within the somatic documentation – the voluminous files from oncology, surgery and radiology – nor in her psychiatric in-patient records.

Four months into this treatment regimen, Cecily attends her routine post-cancer check-up at the hospital's department of radiology. She confides some details about her childhood to the radiologist. This prompts him, at his own initiative and without any official requisition, to take x-rays of her entire body. What they reveal is horrifying: they document a literally broken child. There are x-ray images of 32 fractures – on Cecily's forearms, legs, hands, fingers, toes, face/jaw and thorax – the majority dating from ages 6 to 14. Only the seven fractures that had probably been sustained when Cecily was an adult, the results of sports injuries or a bicycle accident, bear traces of having been treated medically in any appropriate way. The broken child, however, had not received adequate help.

What then did Cecily gradually piece together about her childhood and adolescence?

> *My alcoholic father was unemployed. So my mother held down two jobs and was out of the house from early morning to late at night, every day, all week. My father stayed home, drinking – and playing cards with his brother and some friends a few times a week. I used to clean up when they'd finished so my mother wouldn't come home to a mess. I also took care of my very handicapped younger brother. Every morning before I went to school I'd get him into his wheelchair and to the bus stop, where they picked him up and drove him to the institution where he stayed all day.*
>
> *Afternoons were the worst, when the men were around our kitchen table. That's when I was physically maltreated and sexually abused by all of those more-or-less drunk, repulsive men. They cursed me and beat me. They strong-armed me and pushed and threatened and raped me, vaginally, anally, orally. I was locked in the cellar or tied to a narrow shelf. My brother and I were often starving because my father and his friends ate whatever was in the refrigerator. I had no money to buy food because my father spent whatever was in the house on liquor. I was bullied at school because my clothes were old and worn out. I never brought a proper lunch to school, my hygiene was poor, and I was never taken to a doctor. I couldn't always prevent my brother from being beaten and maltreated as well. I learned to hide my pain, even when these men broke my bones raping or beating me. The worst of them was my uncle. My father never molested me sexually, but in all other ways.*

15.5 Reflections

15.5.1 Recently Acquired Knowledge

The medical context is changing. The history of Cecily Cramer's life and illnesses represents a case-in-point of knowledge that has not simply been accumulating during the last three decades, but also converging, within a broad field of research domains. In 1998, the first findings of the *Adverse Childhood Experience Study* were published in the *American Journal of Preventive Medicine* (Felitti et al. 1998). These documented that childhood hardships, differentiated into ten types, were shown to correlate to adult sickness, in a dose-response relationship; that is, the more hardships experienced during childhood, the higher the risk of becoming ill in later years (Felitti and Anda 2010).

Since then, many similarly designed studies have supported these findings, worldwide. There is now no doubt *that* childhood adversities resembling Cecily's in type and duration are harmful to a person's current and future health and well-being (Shonkoff et al. 2009). The wide spectrum of detrimental effects reaches into every specialty within somatic and psychiatric medicine, as well as impacting the social sectors charged with tackling alcohol and drug abuse, the law enforcement sector, the family welfare agencies that confront divorces, broken homes, partner violence, occupational issues, disabilities. In other words: the economic consequences of child maltreatment and abuse impact a wide range of societal systems and their budgets, constituting a burden that would be heavy for any society to bear (Knudsen et al. 2006).

Once the fact *that* childhood adversity is related to adulthood sickness had been documented, the question then arose as to *how*. How are experiences of strain transformed into pathophysiological processes? An interdisciplinary field of research currently termed *neuroscience* has been developing steadily, step-by-step. It has helped provide a foundation for hypothesising models that might explain the inherent logic of how those processes occur. The neuro-endocrinologists at Rockefeller University, New York, have made a huge contribution: they developed the now widely accepted concept of *allostasis*, a term meaning, "stability through change" (McEwen 1998). The concept of allostasis aids in exploring the flexibility of human physiology, how it may adapt to extreme challenges and maintain viability during long-term strain and hardship. Deriving from allostasis is the concept of *allostatic overload*, referring to what happens when stressful conditions must be endured over excessively long periods of time, with all the energy-providing systems in the body paying the price. When allostatic overload has exhausted the body's flexibility and adaptability, the most central, systemic regulators may break down (Danese and McEwen 2012). Such a "multisystem-dysregulation" is characteristic of complex sickness, although that is still conceptualised and termed "multi-morbidity" (Wiley et al. 2016; Tomasdottir et al. 2015).

The long-term, overwhelming strain of allostatic overload has been shown to take a toll on the immune system and on the hormonal and central nervous systems as well. This means that the regulation of glucose, lipids, minerals, blood pressure, heartbeat, muscular tension, rest, sleep, respiration and digestion are all adversely affected. These systemic disturbances, or overloads, contribute to an array of serious and chronic conditions, such as cardiovascular, respiratory, and liver diseases, as well as type 2 diabetes. They also affect tumour development and the frequency of infections by suppressing the immune system at the cellular level, both those features that are innate and those the system acquires. Simultaneously, the hormonal aspect of the immune system, the inflammatory system, is in a state of constant hyperactivity, engendering systemic inflammatory diseases, among them the so-called autoimmune diseases (Dube et al. 2009; Song et al. 2018). In addition, recent studies indicate a relationship between Post-Traumatic Stress Disorder (PTSD) and neurodegenerative diseases such as Alzheimer's and Parkinson's.

15.5.2 Updating the Concept of Causality

The knowledge that health professionals have now acquired regarding the potentially adverse impact of certain types of lifetime experience on human health **obligates** *them to re-examine the traditional medical understanding of causality.*

For a proper diagnosis to be determined and an adequate treatment offered, the cause of a health problem must be identified correctly; it must not be conflated with aspects and effects of phenomena that are contextual. To paraphrase how a group of scientists at Harvard Medical School put it recently: Whenever the effect of lifetime adversity on the brain is mistakenly labelled as the "cause" of a psychiatric disease, what results is *biased* knowledge (Teicher and Samson 2016; Teicher et al. 2016; Ohashi et al. 2017). Moreover, such "knowledge" masks the actual sources of a variety of types of morbidity and mortality, ones that are apparently intrinsic to the structures of our societies (Farmer et al. 2006; Jones 2000). Treatment based on such mistakes misleads the medical gaze *towards* diseased individuals and groups, and *away from* the pathogenic conditions that they live *with* and *in*. However unintentionally, medicine thus becomes complicit in obscuring abuses of power and all kinds of societal injustice. Must such consequences remain unexplored within medicine because they are defined as lying outside the mandate of the profession?

References and Further Readings

Danese A, McEwen BS (2012) Adverse childhood experiences, allostasis, allostatic load, and age-related disease. A review. Physiol Behav 106:29–39

Dube SR, Fairweather D, Pearson WS, Felitti VJ, Anda RF, Croft JB (2009) Cumulative childhood stress and autoimmune disease in adults. Psychosom Med 71:243–250

Farmer P, Nizeye B, Stulac S, Keshavjee S (2006) Structural violence and clinical medicine. PLoS Med. https://doi.org/10.1371/journal.pmed.0030449

Felitti VJ, Anda RF (2010) The relationship of adverse childhood experiences to adult health, well-being, social function, and health care. In: Lanius R, Vermetten E, Pain C (eds) The effects of early life trauma on health and disease: the hidden epidemic. Cambridge University Press, Cambridge

Felitti VJ, Anda RF, Nordenberg D, Williamson DF, Spitz AM, Edvards V, Koss MP, Marks JS (1998) Relationship of childhood abuse and household dysfunction to many of the leading causes of death in adults. Am J Prev Med 14:245–258

Jones CP (2000) Levels of racism: a theoretical framework and a gardener's tale. Am J Public Health 90:1212–1215

Knudsen EI, Heckman JJ, Cameron JL, Shonkoff JP (2006) Economic, neurobiological, and behavioural perspectives on building America's future workforce. PNAS 103:10155–10162

McEwen BS (1998) Protective and damaging effects of stress mediators. N Engl J Med 338:171–179

Ohashi K, Anderson CM, Bolger EA, Khan A, McGreenery CE, Teicher MH (2017) Childhood maltreatment is associated with alteration in global network fiber-tract architecture independent of history of depression and anxiety. NeuroImage 150:50–59

Shonkoff JP, Boyce WT, McEwen BS (2009) Neuroscience, molecular biology, and the childhood roots of health disparities. Building a new framework for health promotion and disease prevention. J Am Med Assoc 301:2252–2259

Song H, Fang F, Tomasson G, Arnberg FK, Mataix-Cols D, de la Cruz LF, Almquist C, Fall K, Valdimarsdottir UA (2018) Associations of stress-related disorders with subsequent autoimmune disease. J Am Med Assoc 319:2388–2400

Teicher MH, Samson JA (2016) Annual research review: Enduring neurobiological effects of childhood abuse and neglect. J Child Psychol Psychiatry 57:241–266

Teicher MH, Samson JA, Anderson CM, Ohashi K (2016) The effects of childhood maltreatment on brain structure, function and connectivity. Nat Rev Neurosci 17:652–666

Tomasdottir MO, Sigurdsson JA, Petursson H, Kirkengen AL, Krokstad S, McEwen B, Hetlevik I, Getz L (2015) Self-reported childhood difficulties, adult multimorbidity and allostatic load. A cross-sectional analysis of the Norwegian HUNT study. PLoS One. https://doi.org/10.1371/journal.pone.0130591

Wiley JF, Gruenewald TL, Karlamangla AS, Seeman TE (2016) Modeling multisystem physiological dysregulation. Psychosom Med. https://doi.org/10.1097/PSY.000000000000288

Open Access This chapter is licensed under the terms of the Creative Commons Attribution 4.0 International License (http://creativecommons.org/licenses/by/4.0/), which permits use, sharing, adaptation, distribution and reproduction in any medium or format, as long as you give appropriate credit to the original author(s) and the source, provide a link to the Creative Commons license and indicate if changes were made.

The images or other third party material in this chapter are included in the chapter's Creative Commons license, unless indicated otherwise in a credit line to the material. If material is not included in the chapter's Creative Commons license and your intended use is not permitted by statutory regulation or exceeds the permitted use, you will need to obtain permission directly from the copyright holder.

Chapter 16
Conclusion: CauseHealth Recommendations for Making Causal Evidence Clinically Relevant and Informed

Rani Lill Anjum, Samantha Copeland, and Elena Rocca

We have presented an overview of the philosophical framework of CauseHealth and shown how it relates to the clinical encounter. By giving some examples of how the abstract philosophical ideas (Part I) can be implemented into clinical practice (Part II), we hope to have shown how ontological and other foundational considerations are relevant for healthcare professionals. We also hope to have demonstrated that a change toward a genuinely person centred practice cannot happen without a corresponding change in the ontology, norms and methods underlying that practice. Finally, we hope to have empowered and inspired healthcare practitioners to get involved in debates about the foundation of medicine and healthcare and how implicit philosophical assumptions shape and define their profession.

16.1 Practical Recommendations for Change

From the philosophical perspective presented in the first part of this book, taken together with the clinical application of this in the second part, it should now be clear that we must change the way we approach causal evidence of health and illness conceptually, methodologically and practically. This has some practical consequences for the clinical encounter, as discussed throughout the book. We now offer some specific CauseHealth recommendations for making causal evidence more

---

R. L. Anjum (✉) · E. Rocca
NMBU Centre for Applied Philosophy of Science, Norwegian University of Life Sciences, Ås, Norway
e-mail: rani.anjum@nmbu.no; elena.rocca@nmbu.no

S. Copeland
Ethics and Philosophy of Technology Section, Delft University of Technology, Delft, The Netherlands
e-mail: s.m.copeland@tudelft.nl

clinically relevant and better informed by the clinical encounter. The recommendations follow from an overall consideration of the previous chapters, both the philosophical framework and its clinical implications.

Assume medical uniqueness, because there is no normal, standard or statistically average patient.

Individual dispositions and propensities are given by intrinsic properties and their interactions within a unique context. What will happen in such a unique context with a unique set of dispositions can therefore not be derived directly from what happens in similar contexts, for instance in a population of similar individuals. Assuming causal complexity, context-sensitivity and causal singularism, each causal process will be unique, including a unique combination of dispositions. This unique combination of dispositions come from the patient and their total situation. The idea of an average patient is therefore a statistical artefact: in the clinical reality, the standard or average patient does not exist.

Treatment should be adapted rather than standardised, because no two patients are causally equal.

Since causality is intrinsic and unique to a specific context, it follows that an equivalent intervention in two different patients might amount to different medical treatments. Decisions on treatment ought therefore to be tailored to the individual rather than derived from statistical evidence alone. The default expectation is that treatment cannot be standardised, but must instead be adapted to the patient in their specific context, with their unique set of dispositions, aims and preferences at that particular stage of their illness or recovery process. This also means that clinical knowledge and judgement is vital for making decisions about how to proceed with the individual seeking care.

Value qualitative approaches, because causal evidence is much more than evidence from RCTs.

There is no perfect method for establishing causality in the dispositionalist sense; i.e., that the effect was a result of intrinsic dispositions interacting. Instead, we need a plurality of methods and multiple types of evidence. This means that the meaning of 'evidence' needs to be expanded, and not be restricted to evidence from experiments and controlled setups. Qualitative approaches based on clinical dialogue and observation have a huge potential for understanding which causal elements and mechanisms are in place in the individual case. Indeed, qualitative approaches can contribute to study complex causal interactions between multiple factors from biology, social context, medical history, genetics and lifestyle. Causal evidence must also include evidence from the specific patient who is supposed to benefit from the intervention.

Consider mechanistic and theoretical knowledge, because we need to understand hows and whys.

If the healthcare practitioner is to make rational clinical choices for their patients, it is important to consider the plausible causal mechanisms underlying statistical evidence. This means that, for the purpose of the single case, theoretical knowledge about *how* and *why* different causes interact to produce the effect must be given epistemic priority over *how often* an intervention is followed by the effect in a studied population. We must seek to understand causal mechanisms and not be satisfied with evidence from correlation data and comparisons of these. Indeed, knowing what happens to other patients is most useful for clinical decisions for the single patient when we also know why it happened and which dispositional properties were causally relevant for the outcome.

Accept clinical uncertainty, because precise quantitative estimates do not reflect reality.

Causal knowledge and evidence will never be complete, so clinical uncertainty is ineliminable. Any causal process can be interfered with by introducing additional dispositions, and all real-life situations are open systems with an unlimited number of causally relevant factors or dispositions. The lack of certain predictions thus reflects the reality of the clinic. Any prediction made with precise numeric estimates can only come from assuming an idealised model in a deterministic and closed system, using an algorithm to calculate the probability of an outcome. Such predictions might seem certain, but no such certainty can be transferred to practice. Dispositions and propensities do not generate clearly quantifiable predictions, but rather point to qualitative and contextual considerations.

Consider individual propensities, because they affect the risk and safety of treatment.

To understand how an intervention works and which contextual factors might influence the outcome is crucial for assessing potential benefits and harms that the intervention might have for different individuals. This must also include a consideration of long-term effects. For this, we need to consider not only the dispositions of the intervention, but also the unique, individual propensities of the person receiving care and the causally relevant dispositions represented by their life situation. Risk and safety must be considered on the individual level instead of something that can be directly derived from frequencies in populations. Specific care must be taken to pick up on the patient's vulnerability to the intervention.

Know your patient, because most of the causally relevant evidence will come from there.

Causal evidence comes not only from clinical studies and medical research, but also from the patient context. Their biography, genetics, medical history, lifestyle, diet and life situation will represent most of the causally relevant information needed to understand and make predictions in this case. The more we know about the patient, the better we understand their condition and the more causal evidence we have for making good and safe clinical decisions.

Study unexpected outcomes, because there is much to be learned from outliers and marginal cases.

Theoretical knowledge of complex causal interactions cannot easily be gained from controlled experiments alone, where single factors are studied in isolation from background conditions. Only in the single patient will this causal complexity be observed. In marginal and outlier cases, where unexpected or rare outcomes occur, we might be able to observe hitherto unknown causal interactions in a unique combination. Such outlier cases should be studied in detail, since they are an opportunity to learn more about causal mechanisms from rare interactions and contextual interferers.

Clinical evidence should inform research, because that is where causal complexity is observed.

Since the clinical encounter offers us a chance to investigate causal mechanisms in their full real-life complexity, the medical community should make the most out of this opportunity. For instance, when the clinical inquiry is done in a whole person centred way, those results can feed back to basic and clinical research with new causal hypotheses. Information will thus flow from the clinic to research and not only from research to the clinic.

Listen to the story, because 'medically unexplained' does not equal 'no causal explanation'.

That a condition is medically unexplained does not necessarily mean that no causal explanation for the symptoms can be given in the individual case. Sometimes, 'medically unexplained' means simply that none of the causes suspected by the patient or clinician can be backed up by statistical evidence. Although observed repetition would have been a good reason to accept it as a medical cause, causal singularism suggests that no such repetition is required for it to count as a cause, ontologically speaking. The patient's story is a vital source of causal insight into the unique complexity of their condition.

Rebel, because medicine and healthcare must move beyond positivist scientific ideals.

What counts as 'scientific' is too narrowly defined within the current medical paradigm, with emphasis on quantitative data and comparisons of these. Alternatives to these positivist ideals should be welcomed. For instance, clinical reasoning should take into account patient narratives and embodied lived experiences as key sources for investigating causes and effects in the single case. Dispositionalism calls for a radical change in medicine and healthcare, away from reductionism, standardisation, fragmentation and medicalisation, and toward an ecological, phenomenological, whole-ist and genuinely person centred approach.

16.2 Toward a New Paradigm

We are not questioning the idea that medicine and healthcare should be evidence based. On the contrary, we want to challenge the definition of 'evidence', and specifically of 'causal evidence'. From a dispositionalist perspective, what counts as causal evidence is much broader than what is suggested by the current framework. We wish therefore to see the rise of a new paradigm, in which healthcare decisions are not seen as 'evidence based' until they include *all* the causally relevant evidence. This means that we need to consider, not only evidence from general knowledge and research on populations, but crucially also qualitative and phenomenological evidence from the particular encounter with the patient. Downgrading the latter as 'less scientific', 'less reliable', 'anecdotal' or 'secondary', implies an unspoken commitment to a very specific philosophical bias about causation, as we have seen. When translated into clinical practice, those philosophical biases carry an inherent risk of delivering a poorer, de-humanised, fragmented and at times counterproductive healthcare, as the testimonies in this book so powerfully warn.

Open Access This chapter is licensed under the terms of the Creative Commons Attribution 4.0 International License (http://creativecommons.org/licenses/by/4.0/), which permits use, sharing, adaptation, distribution and reproduction in any medium or format, as long as you give appropriate credit to the original author(s) and the source, provide a link to the Creative Commons license and indicate if changes were made.

The images or other third party material in this chapter are included in the chapter's Creative Commons license, unless indicated otherwise in a credit line to the material. If material is not included in the chapter's Creative Commons license and your intended use is not permitted by statutory regulation or exceeds the permitted use, you will need to obtain permission directly from the copyright holder.

Printed by Printforce, the Netherlands